Results from the

Biodynamic Sowing and Planting Calendar

Maria Thun

Results

from the

Biodynamic Sowing and Planting Calendar

Floris Books

First published in German as *Hinweist oens der Konstellationsforschung* by M Thun Verlag
First published in English as *Working on the Land* by Lanthorn Press
This edition translated by Gerhard Staudenmaier from the eighth German edition.

British Library CIP Data available

ISBN 0-86315-420-4

Printed in Poland

Contents

II Soil, Manuring and Composting

III Planting, sowing and harvesting

V Food Preparation

Our Site in Dexbach

Since 1952 we performed trials in Marburg Lahn, Germany. We started on weathered red sandstone soil on a southern slope called 'Blocksberg,' and later added an area of white sandy soil on a northern slope, as well as clay soil near a little stream in a small, sheltered valley, thus giving us different climate conditions. In 1963 we went to Gisselberg in the Lahn Valley, about five kilometres south of Marburg. There was sandy clay soil with a healthy humus layer. All transportation was made by bicycle pulling a trailer.

In Gisselberg we had the chance to collaborate with a biodynamic farmer, Andreas Ortwein, and our trials were conducted within the frame of his crop rotation. For us the work was much easier as he used machines. When he retired he became our co-worker. We found nice ground with sandy clay and a healthy humus layer graded with 65 points and being $^1/_2$ hectare large. We only found stones by chance. In 1967 we started joint trials with Prof. von Boguslawski in Rauischhozhausen, where there were parabrown soil conditions graded with 80 points. Stones could be met only in a depth of more than one metre. Farmers visiting us frequently looked at our magnificent crop fields and asked if our method also would work on their fields with lower grade soil.

The painter, Walter Thun, loved the landscape near Dexbach and Matthias, the beekeeper, realized the importance of this landscape for bees. So, finally we decided to move to Dexbach.

At first we had to break up the wasteland because farmers were not selling any arable land. We tried to plough but down to a depth of five centimetres there were only stones. Another way had to be found. Slowly we tried to enliven the soil. Dr Balzer made the first analysis of the soil and found that it was mostly composed of little stones and had a pH value of 4.5. He needed large amounts to have enough soil for the analysis. Our first co-worker from Dexbach was convinced it would be best to grow potatoes, as they need a lot of air, which was present between the stones. However, he still used every free minute to gather stones. The ground was comprised of weathered slate rock with some huge diabase (Dolerite) in between, which had to be removed because it destroyed the machines. We took it home to border our patches.

Figure 1. Flock of sheep at Dexbach in 1972.

Figure 2. Making the garden paths.

Figure 3: French friends at Thalacker

Figure 4: View from the test fields to Dexbach

Our test fields had soil graded with 80 points and 65 points and wasteland and we performed our trials there for several years. The shepherd put his flock on the wasteland overnight, and we sprayed Manure Concentrate Preparation frequently. We sowed plants for the bees and after flowering they were dug into the soil, and we sprayed Manure Concentrate again.

After a few years farmers from Dexbach village asked us for seed from our plants as they thought the plants were so good looking due to the seed. Also, farmers came again, who thought our soil in Gisselberg was much better than their own. On their way to our test fields they became quieter and quieter and Herbert said: it looks like at home. And he asked: How did you achieve it? We have tried to show this in this book.

The First Nine Years of Research

In his book, *Agriculture,* Rudolf Steiner showed how plant life develops under the influence of the planets. He indicated the positive effects of the forces of full Moon on the growth of seedlings. This idea has been taken up by many biodynamic farmers and gardeners and applied to their daily work. For this reason Franz Rulni and Heinrich Schmidt composed a sowing calendar, which contained the phases and nodes of the Moon, apogee and perigee and, to some extent, the ascending and descending Moon. Applying their advice I made the following observations in 1952.

In one period of the full Moon ten suitable days for sowing were indicated. By chance I sowed radish *(Raphanus sativus radicula)* on four different days, and soon realized differences in growth resulting in totally different plant types. Without delay I wanted to find out the cause for these differences. To try and solve the mystery I made daily sowings.

The trials were started on humus-rich, sandy soil. Soon afterwards the influence on the earlier crops became evident. A new garden was found with sufficient area and the same conditions. The soil was fertilized with mature compost, which had been prepared according to biodynamic compost preparations. Horn Manure Preparation was only sprayed in autumn and spring because there was not enough time for a daily application. Also, because I feared it might have an additional effect on the plants.

At first a particular area was prepared for sowing and then I put one row of seeds into the Earth every day. Soon I realized that the sowings made on the first few days were different, but then a uniform plant type developed.

The new observations could not be explained immediately. Meanwhile another patch planted over a different month was growing, which showed again the same phenomena. A long time of brooding and reflection led me to the assumption that the differences were due to the moment of tending the soil. Therefore only a small strip of Earth was worked on every day and sown, and soon afterwards strong differences were visible.

New problems arose: a drought arose, accompanied by very high temperatures. Following tradition the radish patches were watered, with the consequence that all the rows looked alike, with differences in size due only to later sowing. Only the effect of watering could be seen. This experience led to the decision to leave all plants to the elements. The only work was to hoe and to weed. In this way the trials were continued for several years.

In the meantime I found that root formation occurred rhythmically, so a new question arose: If the full Moon stimulates the part of the plants above ground, can the forces of the new Moon be helpful in root formation? The results of germination trials in water remained ambiguous.

A suitable crop to check this question seemed to be broad beans with their strong roots. So every day I put some broad bean seeds into the soil and observed their growth. A few months later I was convinced that I had found the rhythm — root growth set in on the day when the Moon's node began to descend. Since the rhythms of the Moon and planets were still unknown at that time I was happy about my findings. The summer had gone and the winter impeded observations. New trials began the following year. Soon it became obvious that root formation started one day after the lunar node. The disappointment was large but led finally to the decision to solve the problem, as the rhythm of root formation offered new ways for the grower to save growth time after transplantation.

In 1956, broad beans were sown again. The same results as in the past year were found. However, in autumn the difference between the Moon's node and the beginning of root formation increased, and showed a variation of three days in the following year. This made it clear that any connection with the Moon's node was a false assumption.

It became necessary to study cosmic rhythms in more detail. Finally, I realized that at the beginning of the root formation trials the highest point of the ascending Moon was one day earlier than its node, but later it was on the same day for a certain period. In following years it became two and then three days later owing to the movement of the node along the zodiac. As a result the day found for root formation was a fixed one in the sidereal cycle of the Moon, which was located in the constellation of Gemini at that time, but generally depended on the Spring Equinox. Although the mystery was solved further proof was still needed. So different slips, layers of potted flowers and hyacinths were put into water to observe their root formation over a period of time. Outdoors trials were done with lettuce, cabbage and annual flowers.

Again and again the descending Moon was found to be influential, and its effects intensified by the descending forces of the day. Several things were discovered — plants transplanted in the afternoon, when the Moon was descending, grew without any inhibition of growth if watered only once after planting. This process continued when plants were taken out of the soil in the morning, when still full of the ascending forces of the day. They were kept cool and moist until they could be planted in the afternoon. Overnight the roots adapted to the soil and the next day the plants looked fresh. Compared to the same type of plants that had not been not touched they showed better growth over the whole season. Plants sown at the same time but transplanted during the ascending Moon were stunted for days or weeks. The outer leaves wilted and only with large effort could the plants renew from within and, in practice, one had to water them frequently. Their development was delayed by several weeks.

In the meantime regular plantings of radish were continued taking note of the experiences gained so far. A certain principle was discernible but the ability to read it was still missing. The same plant types developed sometimes for two, three or even four days: for example, a strong root (tuber) but small leaves developed (root type); another one had big leaves and a relatively small root (leaf type); the third type developed leaves on a red stem and hurried to develop stem and flower, but remained in this state for a long time with only a trace of a tuber (flower type); a fourth type seemed to omit tuber formation, and hurried through leaf and flower formation to form seeds to ensure reproduction (fruit/seed type).

The fourfold nature of the plant became evident. At first it was difficult to determine due to the fact that the plant has three organs — root, leaf and flower. The newly-discovered fourfoldness was repeated three times in one month.

Suddenly I found myself confronted with the discovery of a basic law of plant formation. Did the first cultivators have an insight to this law? The knowledge of this fourfold organization was an unexpected reward for many years' work.

On quiet winter days I went over the results trying to explain in detail the diverse formation processes. When the Moon was moving along the zodiac from one constellation to the next the plant type changed. I made a diagram in different colours of the results and realized that I had found a similar structure in the zodiac to one described by Dr Guenther Wachsmuth, on etheric formative forces of the zodiac.

The correspondence is shown here:

Dr Wachsmuth		My observation
Life ether	Taurus	Root formation
Life ether	Virgo	Root formation
Life ether	Capricorn	Root formation
	Earthly forces	
Chemical ether	Pisces	Leaf formation
Chemical ether	Cancer	Leaf formation
Chemical ether	Scorpio	Leaf formation
	Watery forces	
Light ether	Gemini	Flower formation
Light ether	Libra	Flower formation
Light ether	Aquarius	Flower formation
	Light forces	
Warmth/fire ether	Aries	Fruit formation
Warmth/fire ether	Leo	Fruit formation
Warmth/fire ether	Sagittarius	fruit formation
	Warmth forces	

A comparison of the two schemes revealed that certain lunar forces, depending on the Moon's position in the zodiac, stimulated one of the four etheric forces, and affected the plants through the elements, earth, water, air and warmth. In fact the Moon helps to bring the forces of a region of fixed stars to plant formation, at least this was the case for the radishes that had been observed so far.

We then tried to find the reason for the growth inhibitions observed on certain days and during certain periods, and came to the conclusion that perigee always had an unfavourable effect, as did the Moon's node at certain times. Planetary oppositions also changed the plant type. For instance, the opposition of Venus and Jupiter in March 1957 stimulated strong flower development.

Many questions arose but especially this one: What is the effect of these forces on other crops, or do they only affect radishes? We had to examine typical representatives, and in the following years we investigated: root crops (carrots, parsnip and scorzonera); leaf plants (lettuce, spinach, lamb's lettuce, cress and cabbage); flower plants (zinnia, snapdragon and aster); and fruit/seed plants (peas, beans, cucumbers and tomatoes).

To fertilize the plants plant compost was used. In spring and autumn we sprayed Horn Manure Preparation. Horn Silica was not used because it can alter the plant type.

The results for radishes were confirmed for other annual plants. Frequently it was difficult to see the differences in the early stage of growth but by the end the result was unambiguous. Root crops sown with the Moon in Taurus, Virgo and Capricorn developed the best crops being of high quality and tasty, with a relatively small plant above ground. Root crops sown on Leaf days had strong leaf development and a small root and did not keep well. When stored in the cellar they shrivelled in early winter, while root crops sown on Root days kept for much longer and had a salty structure in the spring. Root crops sown on Flower and Fruit days had finely-structured leaves with tough roots, and in 1959, the year of warmth days, some were flowering.

Leafy crops, like spinach, lettuce, lamb's lettuce cress and cabbage, prospered best when sown with the Moon in Pisces, Cancer and Scorpio. The plant developed the largest amount of leaves and remained at the leaf stage for quite a long time without moving on to the flower and seed stages. When the plant was left growing too long it began to develop tillers and this indicated that the reproductive forces were held in the leaf region. This phenomenon was observed more frequently when plants were sown with the Moon in Cancer or Scorpio. In this way we could sow and harvest spinach and lamb's lettuce from spring to autumn without any risk of the plants running to seed. It should be noted that a larger yield of leaves was obtained for sowings made no later than Midsummer's Day in Cancer and Scorpio, and for sowings in Pisces after Midsummer's Day.

For asters, snapdragons and zinnias we observed: Sowings on Leaf days (Pisces, Cancer, Scorpio) led to a huge growth of leaves but the plant had difficulties forming flowers. Sowings on Fruit days (Aries, Leo, Sagittarius) prospered sparsely, had very few leaves and quickly developed flowers and then seeds. The flower plants flourished best when sown with the Moon in Gemini, Libra and Aquarius. Plants developed many branches and numerous flower buds and the plant remained quite long at the flowering stage. If we cut flowers the plant put forth new side shoots and the flowering activity was remarkable. Flower plants sown on Root days had miserable growth above ground and very few flowers.

The observation of fruit plants confronted us with new mysteries. It turned out that Aries and Sagittarius were favourable for the development of the flesh of cucumbers, tomatoes and beans, but were not favourable for peas and broad beans. Peas and broad beans sown when the Moon was in Leo led to the best harvest. Sowings of cucumbers, tomatoes and beans developed very rapidly initially and formed small fruit with big seeds. At first I assumed that this was the effect of a new influence and performed new trials. We had found a fourfold pattern, was there something else now?

Cucumbers, tomatoes, beans and peas sown on Leaf days behaved as shown already: with a huge growth of leaves and lazy flowering. We observed several times that cucumbers sown on flower days had a lot of male flowers and few fruit-bearing female flowers. The plant was only able to reproduce itself, which was not satisfactory for the grower. Cucumbers, tomatoes and beans yielded excellent fruit flesh when sown with the Moon in Aries and Sagittarius but Root day sowings developed only scarce growth above ground; frequently growth inhibition took place followed by attacks of greenfly. Sowings of runner beans when the Moon was descending in Virgo led to a large amount of roots at considerable depths; roots were thick like a turnip and composed mostly of nitrogen-containing root nodules. Similar observations were made for peas and broad beans. It seemed that these days were most suitable for sowing plants with a large amount of roots intended for green manuring.

From numerous sowings of fruit and seed plants with the Moon in Leo, Sagittarius and Aries we finally obtained the clear result that these forces subdivide into Sagittarius-Aries days, which promote fruit flesh, and Leo days, which stimulate seed formation. So, we found a fivefold nature.

In 1962, when I was looking through the lectures of Rudolf Steiner,* I was surprised to find by chance a fivefold nature of plants and their therapeutic effect on the human being. He showed the healing forces of seeds on the human heart, and said that the forces forming the region of the heart in the physical body originate in the constellation of Leo. It is a very cheering feeling to have observed that a heart disease plant remedy is stimulated by the same constellation.

* Lecture of January 1, 1912. *The World of the Senses and the World of the Spirit*, Steiner Book Centre 1979.

These findings encouraged me to publish the results of my work although I am aware that a lot of things still need to be completed. We need to study bi-annual plants, cereals, trees and shrubs to find other rhythms. However, trees and shrubs are not within the field of my expertise.

The work reported so far also required detailed observations of the weather and the micro-climate, which was influenced by the location of the fields, dew formation, breathing of the Earth, its way of retaining warmth etc. We found that the movement of the planets along the zodiac determined the general weather situation. Intermediate irregularities were due to conjunctions and oppositions. In the past earthquakes occurred at the same time as squares with Neptune, and strong thunderstorms with squares of Uranus. However, the micro-climate correlated with the Moon's position in the zodiac. Dew and mist were different on watery or warmth days. When the Moon was in Pisces, Cancer and Scorpio the sky was frequently grey and in combination with the wider weather conditions those days often had steady rain. When the planets did not support humidity a layer of haze was present on such days giving the soil humidity. This was not present on other days. This humidity had to be captured by working the soil in dry periods. We observed that during moist and cool weather the atmosphere was different on warmth and light days compared with watery and Earth days. Working the soil gave the possibility to bring either warmth or air into it, and to keep or free the humidity.

In the first years of research we observed that working the soil prior to sowing had a distinct effect on the plant's development. Now a new question arose: Was it possible to stimulate a plant type by working the soil on a corresponding day?

All results reported so far were first obtained on sandy humus soil and then on loam. The soil was tended in the best biodynamic way. Nevertheless, over the course of the year we realized that plants became more and more sensitive to cosmic rhythms. After having gained an overall picture the results were given to other people. I concluded, after personal checks, that mineralized soil hardly reacted to fine cosmic forces, while humus soil had the ability to mediate these forces to the plants. Often we have been asked: What is an enlivened soil? Now it was obvious. It is a soil fertilized with mature compost, which has the ability to absorb cosmic forces and to mediate them to plant-formation processes, thus enabling the plants to activate starry forces, so creating a new quality of human food.

PART ONE

Astronomy

Cosmic Rhythms and Constellations

All life patterns of the natural kingdom — plants, animals and human beings — are woven into cosmic rhythms. It is difficult for people of our time, who consider themselves objective, to accept such a fact. In their subconscious they still live predominantly at a time when the Earth was the unshakable centre around which all the stars moved. How could it be otherwise since when speaking of the effects of the stars on plant growth people say: "How can what happens above have an effect down here on Earth? I don't believe that!" When one hears such frequently-expressed doubts it is clear that people cannot conceive of the Earth participating in the planetary rhythms of our solar system, which has a rhythm of its own, but is also connected with planetary and solar rhythms as they interrelate with life on Earth. After all, planetary rhythms and constellations do not have their origins in theology or psychiatry but are scientific, astronomical facts which one can get to know like any other field of knowledge.

By rhythms we understand certain sequences which arise through movement and which are linked to fixed points. If, for example, we observe a planet moving past a certain fixed star and then follow its path until it reaches this fixed star again we see that it has completed its orbit — its *sidereal* rhythm. In space we have ascertained its relationship to time.

By the same token we see a different situation with the constellations; namely the connections of the planets to each other or to the Sun. We can study, for example, the phases of illumination that can be observed in the relationship of the Moon, Venus and Mercury to the Sun, when viewed from our own standpoint on the Earth. They appear as angles, or aspects, which result from the distance of the planets from each other, in relation to the Earth.

Rhythms are related to the forces of the universe which pour in, and then work through, the layers around the Earth, and find their way in to the Earth by means of the four traditional elements of warmth (fire), air/light, water and earth. Constellation influences also take this path to a certain extent insofar as one is speaking of the "traditional" planets. In addition, there are those planets that only entered human consciousness in the last few hundred years through the use of telescopes, namely Uranus, Neptune and Pluto. Their influences are reflected outwards via the central forces of the Earth itself in the form of electricity, magnetism, volcanic activity and the like. These opposite directions of force sometimes weaken each other, partially overlap in their effect, or even obliterate each other.

Earth, Sun and Moon: Sidereal and Synodic Relationships

Human life as well as animal and plant life is strongly dependent on the rhythms of the Earth. As the Earth turns on its axis over the course of twenty-four hours different parts are illuminated in turn by the Sun, on average for about twelve hours. For human beings this means that they live in *day consciousness* for this period of time, because only with artificial sources of light can the boundaries be extended. The other side of the Earth faces the starry night sky, which implies the condition of sleep. In this way we can speak of people being connected to the *day consciousness* of the Sun and the *night consciousnes*s of the stars. Day consciousness wears down our bodily organism whereas in sleep during our night the damage and devitalization of the day are made good and healed.

In the same way the Earth itself experiences rhythmic change. When turned towards the Sun the plant's light metabolism is stimulated, but-during night-time it is immersed much more in a consciousness of the stars and is stimulated into growth. This means that the night brings vitality whereas the day's experience tends towards the opposite.

Two cosmic aspects of the Earth run in parallel. Its sidereal rhythm, which is an opening to the fixed stars, can be better observed at night. The whole period in which the Earth rotates to face the same point of the zodiac consists of 23 hours, 56 minutes and 4 seconds — the *sidereal* day. The connection with the Sun is different. The *solar* day lasts on average 24 hours with deviations of up to 28 seconds. It is therefore longer than the sidereal day. The year has 365 Sun days while in the same period of time the fixed stars rise and set 366 times (*see* Schultz, *Movements and Rhythms of the Stars*).

Let us now follow the Earth over the course of the year. It moves on its path round the Sun (the ecliptic) in 365 days, 6 hours and 9 minutes. Our year has 365 days. Every four years an additional day is fitted in so that the difference of the six hours can be regulated. A further consideration is that every twenty-five leap years, that is a century, one day must be excluded so that the nine minutes can be balanced out. It is typical of rhythms that they do not follow mechanical clock time.

While the Earth is moving round the Sun during the year the Sun appears in front of the twelve regions of the zodiac, obscuring the stars with its light. When by day the Sun is shining in front of the region of Aquarius then the Earth is facing Leo at midnight — in other words the opposite constellation of the zodiac.

The seasons

The Earth's revolution around the Sun determines the season. For the axis of the Earth's daily rotation is inclined by 23° to the plane of annual revolution. Thus in summer the Sun appears higher in the sky at noon than in winter. The length of the day is longer when the Sun is higher. Towards the poles of the Earth the days lengthen to 24 hours in summer and continual darkness in winter. While at the equator throughout the year each day is 12 hours and each night is 12 hours. Of course in the southern hemisphere the seasons are opposite those of the north, with summer being in December/January. Another rhythm of the seasons is caused by the Earth's elliptical orbit around the Sun. In January it is closest to the Sun (perihelion) and in July furthest (ahelion).

However, the axis of the Earth's daily rotation does not remain stationary. It slowly moves by 1° every 72 years against the background of the stars, causing the celestial equator's crossing point with the ecliptic to move against the zodiac. This is called the precession of the equinox. The equinox will move by 30° (the average size of a constellation) in about 2160 years, thus taking about 26,000 years to move right round the zodiac. (This period is also called the Platonic year).

Two thousand years ago the equinox was in the constellation of Aries, while at present it is in the constellation of Pisces, and in future it will move into Aquarius. The direction of the equinox movement is opposite to that of the Sun's annual path, and opposite the movement of Moon and planets through the zodiac.

The astrological equal-sized *signs* of 30° always begin with Aries on March 21, while the visible astronomical *constellations* of varying sizes are now "out of phase" by about 1 sign, as the Sun is in the *constellation* of Pisces on March 21. It is vital to recognize this difference as all the references in the Sowing Calendar are to the constellations.

Plant cultivation through the year

In his agricultural course Rudolf Steiner pointed to the fact that the Sun has a different quality depending on the constellation in front of which it moves. One should really speak of a Taurus-Sun, a Leo-Sun, and so on. It became clear to us that at various sowing times the whole gesture of the plant changed just as the Sun was entering another constellation. From Earth we can experience an orientation of the stars towards the zodiac. In terms of planting one speaks of plants' compatibility with early or late sowings. We discovered in our experiments that oats thrive best when sown around the time when the Sun is in Aquarius and the Moon is in a constellation which encourages seed formation — in other words between 16 February and 10 March. This brings good growth and good grain yields. If oats are sown with the Sun in Pisces there is stronger leaf development and a tendency towards fungal attack. This time is more favourable for certain root crops. If one sows field beans when the Sun is in the Light region of Aquarius, with its favourable impulses for seed formation, they grow well without incurring aphids. However, if one sows with Sun in Pisces one must reckon on attack by pests because the leaf development has been over-stimulated. The tomato ripens best of all when sown with the Sun in Aquarius whereas the period of the Sun in Pisces brings fungal attacks and a slowing down of the ripening process. If one sows celeriac with Sun in Aquarius there is always a tendency for the plants to bolt. All they need is one cool night in June and the formation of the roots is harmed.

However, this is not the only orientation of the Sun to the zodiac. We have spoken of a solar day and a sidereal day from the perspective of the Earth. However, the Sun too faces different parts of the zodiac. In the space of about 27 days it turns on its own axis, facing each of the twelve regions of the zodiac in turn, rotating through about 13° per day. This is really the Sun's orientation to the stars. One cannot perceive the Sun's rotation with the naked eye. But there is a luminary which helps us to read this orientation to the stars, and this is the Moon. Its circuit round the Earth in 27 days marks out very clearly the twelve regions of the zodiac, from which great creative forces flow. The Moon does not rotate like the Sun and the Earth, but presents the same side towards the Earth, its far side always facing away into space.

Rudolf Steiner speaks of the soil as a living organism in which clay processes must work in order that the upper and lower planetary forces can be assimilated by the plant. Mediating Sun forces come to expression in the clay. In our experiments we discovered again and again that cosmic influences work within the soil. This soil with its clay-humus complex must be well broken down. Here it seems that the Sun's sidereal orientation creates the preconditions for the sidereal Moon to provide the right impulses for plant growth. Perhaps one can understand something Johannes Kepler once said:

> The soul of the Earth seems to be a kind of flame giving subterranean warmth: there is no reproduction without warmth. A certain image of the zodiac and of the whole firmament has been stamped by God into the soul of the Earth. This is the link between the heavenly and the earthly forces and the cause of sympathy between them. The archetypal forms of all their movements and activities have been planted into them by God their creator.

A scientist with such inner piety as Kepler could feel and speak in this way. This can stimulate us not only to plan exact experiments but also to practise good husbandry.

The Zodiac and the Elements

The zodiac, as it has been called since ancient times, is the belt of stars in the sky in front of which the ecliptic runs, and in front of which all the planets of our solar system move on their paths. The influences of other groups of stars are difficult to gauge since there are never any planets passing them. The influences or radiations of the zodiacal constellations are disturbed by the passing planets, which results in an interruption, weakening or strengthening of their influence, which is reflected in plant growth or in weather formation. This can be seen best of all in connection with lunar rhythms, since the Moon as satellite is closest to the Earth and completes its orbit in the shortest time. In experiments with plants and in weather formation one therefore comes across many repetitions. As we mentioned before, the Moon's orbit takes twenty-seven days. Jupiter takes twelve years; Saturn thirty years; and Uranus eighty-four years. How in these cases can one do experiments? Thus the Moon is the best object for experiments and observations.

When the Moon travels past the different constellations on its orbit round the Earth we find:

Pisces, Cancer and Scorpio have a tendency to the watery element.
Aries, Leo and Sagittarius have a tendency to the warmth element.
Taurus, Virgo and Capricorn have a tendency to the Earth element (cold tendency).
Gemini, Libra and Aquarius have a tendency to the airy and light element.

These observations carried out over many years have been confirmed by weather diagrams, collected at the Waldorf School, Engelberg, Germany, by Suso Vetter. These extended over a seven-year period. Eleven types of aspects were evaluated according to coldwarm, dry-damp, high and low pressure in connection with weather observations.

Constellations and signs of the zodiac

Since ancient times people have given the name 'zodiac' to that band of stars across which the planets and the Moon move on their paths. And if we observe the heavens we find the planets move only in front of these twelve constellations.

In the northern hemisphere the beginning of spring comes on March 20/21, which we also call the Spring Equinox. This situation occurs when the Sun crosses the celestial equator. The crossing point of the Sun's path (the ecliptic) and the celestial equator is the vernal point from which degrees of celestial longtitude are calculated. At present the Sun leaves the constellation of Aquarius on March 10 and moves into Pisces. Every day it moves approximately one degree, and on its tenth day in Pisces it arrives at the vernal point.

The vernal point does not remain still, but very slowly moves in the opposite direction of the Sun's annual movement. It takes 72 years for the vernal point to move 1°, so that in 2160 years it moves 30°. Thus through the ages the visible constellations appear at different times of year. The visible *constellations* vary in size — during the course of the year the Sun takes about 6 weeks to traverse the large constellation of the Virgin, while remaining only 3 weeks in the Crab.

Astrologers use a different system — the zodiac is divided into twelve equal segments of 30° (or 'months') beginning with the vernal equinox. These equal sections of 30° are called *signs* and this always begins in March 20/21, regardless of which constellation is visible there in different historical epochs.

Confusingly both astrologers (using regular *signs*) and astronomers (using different-sized *constellations*) use the same Latin (or English) names.

Our research has found that the effects in plant growth correspond to the astronomical *constellations*.

As the length of the constellations differs, we find the impulses affecting plant growth vary from one-and-a-half to four days, as the Moon passes in front of them. For example, if we see the Moon in the *constellation* of Virgo it may already be in the *sign* Libra or Scorpio. However, the plant receives the 'Virgo impulse' through the Moon which stimulates the formation of good roots. Once it moves into the constellation of Libra it begins to transmit forces that stimulate the flower, and, in the constellation of Scorpio the growth forces of the leaves are activated. Our research results show repeatedly that both plants and animals live in accordance with the movements through the visible constellations.

In Figure 5 one can see that there are times when the constellations and the signs partly overlap so that on some days traditional farming lore holds true, but the majority does not. In light of this one can understand a remark by Rudolf Steiner made in 1924 in *Agriculture*: 'A science of these things does not yet exist; people are not yet willing enough to set to work and find it.' Now this comment is no longer quite true since the recommendations in the *Sowing and Planting Calendar* are based on the research since that time. Different results in other places may be due to low humus content or un-rotted manure in the soil. In the case of one farmer who had not used pigs' bristles as manure for eight years, one still found un-rotted bunches after ploughing. They need to be composted until completely decomposed otherwise they block the influences of distant cosmic forces. There have been other similar cases.

Figure 5.

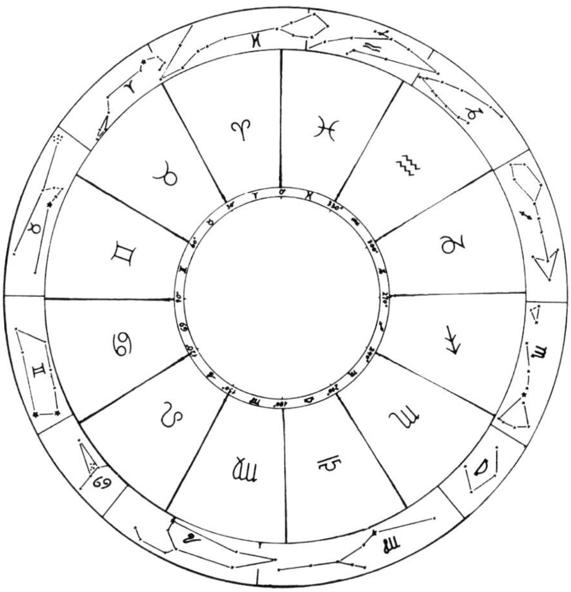

The Moon

The sidereal Moon

We will first direct our attention to the sidereal rhythm of the Moon. Cultivated plants that do not become woody live in close relationship with this rhythm, both with regard to their growth forces as well as their ability to develop foodstuffs in their various organs —root, leaf, flower, fruit and seed. Long observation has shown that forces coming from the fixed stars beyond the Moon's orbit work in various ways upon the Earth and within the soil, and thus also influence the plant. The different regions of the zodiac engender favourable conditions for the development of particular plant organs as the Moon passes in front of the particular constellation — that is, if cultivation, sowing and planting are carried out at the corresponding period. These effects are differentiated in the following manner:

Moon in Taurus, Virgo and Capricorn	— Root development
Moon in Gemini, Libra and Aquarius	— Formation of the Flower
Moon in Cancer, Scorpio and Pisces	— Leaf region
Moon in Leo, Sagittarius and Aries	— Fruit/Seed region

The last group shows clearly that as the Moon passes through Leo, not only is the formation of fruit and seed furthered, but we also find that under this influence the seed quality is definitely enhanced. We experience four formative trends which appear in the sequence root, flower, leaf and fruit/seed, and which are repeated three times in the course of twenty-seven days. The period of time during which each impulse is active varies in length, from between one and a half to four days. As regards the health and yield of the plants, there is hardly any difference to be found between the three related impulses. The inner quality, however, is individual to each constellation; it seems that here the Moon becomes the reflector of the ever-changing quality of the Sun throughout the course of the year, and this fact can at times be observed in the analyses.

Perigee and apogee of the Moon

Since the Moon's orbit round the Earth is elliptical, its distance from the Earth is not always the same. The Moon landings, for instance, are always undertaken when the Moon is nearest to the Earth, for then it is closer to the Earth by 40,000 km than when it is at apogee.

When the Moon recedes from the Earth during its monthly cycle, the effect on plant growth can in some ways be compared with that time of year when the Earth is furthest away from the Sun, i.e. midsummer. The tendency in the plant world is then to run to seed, whereas the growth forces decrease. Thus the effect of the Moon's apogee on seed plants can still be comparatively beneficial. For the sowing of leaf crops, however, this time is definitely unfavourable. Carrots sown during this time easily become woody. The only plant to react positively to being planted at apogee is the potato.

The Moon's perigee, which can be compared to midwinter when the Earth is nearer to the Sun, has a very different effect. If we prepare a seed bed on this day and sow our seeds, germination is poor. Most of these plants are somewhat inhibited in their growth and are also more subject to attacks from fungus diseases and pests. Apogee days are mainly clear and bright, while those at perigee are mostly overcast, rainy, or close. Generally these effects last for a day, but if perigee is in a Root constellation the effect lasts for two days.

Plant life develops in the harmonious interplay between Earth and Sun. Through its roots the plant is drawn down into the earthly realm, while above the soil it gives itself up to the Sun. This harmonious balance is altered through the forces of the Moon at perigee and apogee. This can also be clearly seen in follow-up experiments; plants sown at apogee are drawn away from their earthly hold, while those sown at perigee do not properly manage to place themselves within their own Sun impulse. The following question arose, which had to be solved by new experiments: How far can these hindrances be overcome by applying Horn Manure (500) and Horn Silica (501) Preparations? Previous experiences confirmed that biodynamic sprays do, in fact, strengthen the cosmic effect of any particular day.

Nodes and eclipses

Other aspects that recur rhythmically and strongly influence plant growth are the nodes. While the Moon moves through the zodiac it does not move exactly on the ecliptic (the Sun's path) but on a path inclined to this by 5°. Where its path crosses the ecliptic they intersect and this point is called a node. If the Moon crosses the node at New Moon, it will stand exactly in front of the Sun thus causing a solar eclipse. On the other hand, if it is Full Moon when the Moon is at a node, (and the Sun is at the opposite node), the shadow of the Earth falls on the Moon and a lunar eclipse occurs.

Sowings as well as plantings which were made during these hours often produced variations in the habit of the plants; indeed, even when sowings are made with only one of the planets or the Moon at the node, it is likely that future growth will be adversely affected. It appears that the effect, which these intersections or nodes have, make it advisable to avoid these particular times when working with plants. They have, therefore, been taken into account in the compilation of the *Sowing and Planting Calendar*. Repeated observations have shown that certain plants were strongly inhibited in their development, for example, those which had been sown on days when Mercury, Venus or Mars were crossing the ecliptic, or on days when the Moon was obscuring other planets (referred to astronomically as occultations). The effects can be noticed partly during the planting season, but are often more noticeable the following year. The worst result is a serious decline in the quality of the seed, even going so far as a breakdown of the regenerative powers. Occultations of Uranus repeatedly had these extreme effects.

The observations and experience in connection with solar and lunar eclipses were followed up by further experiments investigating the same kind of constellation, but in respect of other planets. These confirmed again and again that any planetary occultations — or conjunctions, which come close to being occultations — have a similar effect as eclipses or nodal days. We were always on the lookout for disturbances, which were as yet unexplained, and found new aspects, which had not yet been tested. For instance, when a lunar eclipse occurs two weeks after a solar eclipse (when the Moon is in opposition to the Sun at the opposite node), or when it precedes it by two weeks. When any two planets are on the same plane, with the Earth between them, the same kind of constellation can occur between them. This means, however, that there is not only an interruption of planetary influence when the planet is covered by another planet, but that the effect is also interrupted when the planet approaches the node in opposition and the Earth is between them. However, it is not clear whether this can also be regarded as a direct effect of the Earth's shadow. All the same, it should be emphasized once more that there is definitely a loss of force which can be observed when cultivation, sowing and planting are carried out on these particular days. In the same way positive cosmic forces are active at other times, and stimulate plant growth, improve health and increase the yield.

The ascending and the descending Moon

These rhythms must not be confused with the proximity or distance of the Moon nor with the waxing or waning of the Moon. In order to understand this better let us get some help from the Sun's yearly course. Around Christmas time the Sun is at its lowest point and stands in the constellation of Sagittarius. It traces a very flat arc over the southern sky. The point where it rises is in the south-east and it sets in the south-west. In the New Year it soon begins to rise again. Every daily arc rises a little higher in the south. Every morning the point of rising moves more towards the east, and its setting towards the west. The Sun remains longer above the horizon so the days get longer. At the equinox there are twelve hours of day and twelve hours of night. When the Sun is mid-way in the southern sky it is the beginning of spring. Only at St John's (the beginning of Summer) does the Sun reach its highest point in the region of Gemini. This means that the Sun has been climbing all this time up to its highest point. The point where it rises moves more to the north-east. It moves in a wide arc across the sky and sets in the north-west. The Sun has been ascending in the first half of the year and during the second half the arcs descend until the Sun reaches its lowest point at Christmas time in the region of Sagittarius (this description refers to the northern hemisphere).

This is not only the pattern set by the Sun. All planets follow this pattern. During one half of their orbit they ascend and in the other half they descend. Saturn has an orbit of thirty years, fifteen of which are ascending and fifteen descending. Jupiter's twelve-year orbit has it ascending for six and descending for another six years. Whenever planets move in front of the constellations of Sagittarius, Capricorn, Aquarius, Pisces, Aries and Taurus they are ascending. On the other hand when they are moving through the regions of Gemini, Cancer, Leo, Virgo, Libra and Scorpio they are descending. The points of rising and setting move in a corresponding way to the heights of the paths, which can be seen in the diagram below.

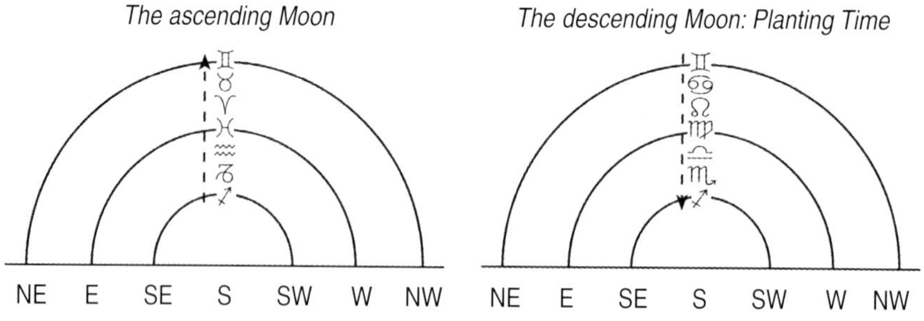

Figure 6.

The course of the day

As we showed earlier the ascending and descending forces are also experienced during a day. The hours before midday work with the ascending Moon forces while the afternoons and evenings involve the descending Moon. However one should exclude the midday and midnight hours.

Phases of the Moon

The synodic Moon cycle refers to the phases of the Moon's which are determined by the relative position of Sun and Moon. We see the same phase every 29.6 days. If the Moon is in the same direction as the Sun —that is, in conjunction — then we have New Moon, when the far side of the Moon is illuminated by the Sun. During the following fourteen days the illuminated part increases and after a little over two weeks it is in opposition to the Sun — in other words it is Full Moon. After this the illuminated part decreases and disappears from our sight again as New Moon. At the end of 27 days the Moon has orbited the Earth once but it is not yet Full Moon, for in the meantime the Sun has moved on approximately 27°. The Moon then needs another two days to reach opposition with the Sun again. Each Full Moon is usually in a different constellation of the zodiac and for this reason it is difficult to compare them with each other. One would have to take the same Full Moon in the same zodiacal constellation over a period of many years for experimental purposes.

While Full Moon is visible all night, New Moon always moves across the sky with the Sun. Each conjunction is almost always in another constellation, causing the same difficulties. In his agricultural course Rudolf Steiner pointed in many different ways to the forces at Full Moon, above all to the way they work through the watery element. What significance do all these phenomena have for plant growth? A certain amount of research has been carried out into the way sea creatures reproduce themselves at certain phases of the Moon. However, much that is attributed to the Full Moon is linked with other rhythms. If one looks at traditional folklore one can find the same influences ascribed to the waxing Moon in Switzerland are ascribed to the ascending Moon in Austria (a direct neighbour). There are many such examples. It is impossible to find a clear understanding in these traditions.

In his course Rudolf Steiner points to the Moon forces working via water and describes how ancient Indians sowed their crops according to the phases of the Moon. If we look at rice, India's main cereal, we are dealing with a culture based on water. Rudolf Steiner ends this indication with the words: "There is not yet a science about these things; no one wants to make the effort to develop it."

Lili Kolisko took up this challenge and, as a researcher, worked on these questions. She carried out sowings of wheat in a laboratory over a period of seven years, comparing the effects of the New and Full Moon. For her evaluation she compared the first two leaves. As far as yields were concerned the results showed a more favourable influence from the Full Moon. If one takes her yearly graph into consideration it is clear that this reflects the course of the Sun. In the middle of summer when no sowings of wheat are made the yields of leaves are at their highest and the differences between Full and New Moon are at their greatest. In spring and autumn when sowings are carried out in the open the yields are low and the differences between New and Full Moon are small. Unfortunately the plants were not allowed to bear seeds so one cannot glean anything of more practical use. The results were published in 1933. There are further descriptions of experiments with vegetable plants carried out in the open over two years. The Full Moon always produces higher yields than the New Moon. It is unfortunate that only the phenomena were described and that there are no figures for yields. From her whole work one feels that Lili Kolisko was at pains to show the connection between lunar influences and water. We quote here the summary of her seven years of research.

There are two extreme possibilities:
a) Too strong a Full Moon influence causes fruit to rot.
b) Too strong a New Moon influence causes fruit to become woody.

> "The Moon forces are guided on the paths of water." Rudolf Steiner said this in 1924 and if one studies the relationship of water and the Moon one begins to understand this indication. If there is dry soil at Full Moon then the lunar forces cannot work properly in the plant because of the missing medium that leads the Moon forces into the plants. If it rains at New Moon the plants do indeed grow, but the New Moon cannot assimilate the watery element into plant growth as the Full Moon can. It is not merely a question of whether the soil is dry or damp, but whether the forces of the Full or the New Moon are linked with the watery element.
>
> The quarters of the Moon occupy an intermediary place as we saw in many experiments.
>
> When there is an exact science of the relationship between the Moon and the watery element on Earth, then one will have the possibility of carrying out sowings for every plant so that the most favourable results can be achieved. With these statements we believe we have made a small contribution to the general problem: the Moon and Plant Growth. Experiments will be continued partly with the same plants and partly with new plants.

In the Sowing Calendar of 1978 we stated:

> In experiments carried out over many years the Full Moon only brought higher yields when a strong manure of mineral origin was applied or not rotted organic manure rich in nitrogen. In both cases the quality was inadequate as far as storage or germination were concerned.

After four years of further experiments we confirmed this. We can only gather such knowledge when we compare storage quality, and use a series of seeds with sufficient variants. In later sowings certain adverse symptoms emerged with the Full Moon seed if the cultures were watered. It appears to me that complaints about the storage quality of crops can be traced back to this.

In conclusion one can say: if our plants grow out of the enlivened soil that we strive for in biodynamic agriculture then the Full Moon influence does not stand out. If we use the wrong manures or water at the wrong times, it appears in the plant growth and brings greater yields of lesser quality. One can observe the same phenomena when seeds are watered to bring about quicker germination.

We can also say that the connections of lunar phases, which we find among the lower animals in the realm of reproduction, appear again in the plant when there is too strong an orientation to the watery element, or when nitrogen has not been broken down in the soil by soil organisms. If humus is in a state of equilibrium and if the values of the humus are not below two per cent, then the zodiacal influences we spoke of in the introduction can affect the plant via the soil. It is then that the sidereal Moon will mediate such influences and become a timer for them.

The rotation of Sun and Moon

The Moon really belongs to the Earth, and although it is true that it has withdrawn to a position beyond the Earth, it can only move a certain distance away from it. It gazes at the Earth with a kind of magic spell, always presenting the same face as if not daring to turn round. Rhythmically it makes an attempt to free itself of the Earth. During each circuit of the Earth it draws further away from its average distance (apogee), but then compensates about 2 weeks later by coming closer (perigee). Plants experience this when they are sown at the time of apogee. If one repeatedly sows at this time over many years then the plants become like climbing plants, and their length is exaggerated. On the other hand, it becomes a catastrophe if sowing takes place when the Moon is very close to the Earth for the plant world which is part of the Earth. Plants then appear as if pushed into the Earth and cannot properly unfold their upper parts. They are attacked by fungus and do not develop healthy seeds. On the Moon itself on such a day there are Moonquakes. When Perigee takes place with the Moon moving in front of constellations, which have a connection with the Earth element, then the phenomena are particularly intense.

All the planets of our solar system have their own rotation, even the Sun, which from a certain point of view is a fixed star, a glowing ball of gas. In the case of the Sun we are not dealing with solid matter, as can be seen by the fact that the middle (or equator) of the Sun does not rotate at the same speed as the parts further away. The differences lie between 23 and 27 days. This means that the surface of the Sun faces the whole of the surrounding cosmos, the zodiac, once over a period of about 27 days. This rhythm is the same as that of the Moon through the zodiac, and we can see the Moon as a kind of 'pointer' to the rhythm of the Sun's facing the different parts of the zodiac.The Moon becomes a kind of transmitter of cosmic impulses to which the plant reacts, as do soil organisms and the four traditional elements, which then brings about the microclimate. The plant receives impulses to develop its different fruiting organs.

The Planets

Planets are luminaries which rotate on their own axes, and can therefore face in all directions. We experience this very clearly with the planet Earth because day and night arise as a result of its rotation. Besides this rotation all planets move round the Sun. We experience this on the Earth too because due to the Earth's path round the Sun the course of the year changes, and as a result of the inclination of the Earth's axis the seasons occur. However, the weather during the seasons is very changeable over a time span of several years and undergoes great fluctuations. These are determined by planetary rhythms, which have different effects on the Earth because of their changed positions.

When one looks at diagrams of planetary movements one has the impression that they are part of a great round dance with many participants — they dance around the Sun. Seen from the Earth they sometimes move directly from west to east behind the Sun, then they move in loops, before resuming their orbit round the Sun. In the case of Mercury and Venus we can observe this clearly. Mercury moves round the Sun three times a year. Every 116 days it finds itself at the closest point to the Earth, and from there exerts its strongest influence on the Earth.

When we observe Venus in front of the Sun it forms similar loops. It swings round the Sun in a much wider orbit, makes a loop in front of the Sun and after 584 days again arrives at its closest point to the Earth, between the Sun and the Earth.

With Mars it is different. For one year it goes behind the Sun and moves with the Sun during the day. It then tears itself away and hurries to the other side of the Earth and from there makes a loop as if in homage to the Sun. During this time it glows red at night, reflecting the light of the Sun, and exerting its strongest influence on the Earth.

The other planets are different again. They move more peacefully on their paths beyond the Sun and the Earth, and enclose what is within their orbits. When they move behind the Sun they move sedately and evenly. When they are on the side of the Earth and come closer to the Earth they move into opposition with the Sun, making a loop by moving 'backwards' as if with a bow to the Sun as in a round dance (possibly there is also a bow to the Earth). We can observe that the strongest influence on the Earth occurs when all the planets make these loops. In the case of Jupiter, Saturn, Uranus, Neptune and Pluto there is one annual loop, with a bow during the time of opposition to the Sun.

When conjunctions occur, the planets are at different distances from the Earth, one behind the other. We then notice that they do not exert their full influence — one hinders the other, and they are prevented from exerting their full forces. If we attempt, as did the ancient Greeks, to see a divine being in each planet, then we would discover different forms of expression of these beings during the various constellations.

In a planet's individual rotation one might see the awakening of a self-awareness; while in a conjunction a resigned abandonment of its own influence, even to the point of virtual paralysis in its activity.

During *opposition* self-aware beings gaze at one another, and in the meeting of their gaze the power of the individual is heightened and works in a creative way on the Earth.

When they come into a *trine* with 120° between the planets (when viewed from the Earth), they form intermediate situations in which higher zodiacal forces can reveal themselves. These influences almost bear an imprint of the Trinity as was experienced in original Christianity. One is then reminded of passages in Genesis when it says, "The spirit of God hovered over the waters."

With a *quadrature* (a 90° distance between two planets) the planets appear to search for the forces they have left behind on the Earth. In the case of the traditional planets one is dealing with forces working inwards from the periphery, which are somewhat lessened and harmless. With the three outer planets it is different since they have their allies in the central forces region — that is, forces working outwards from the Earth — and during quadratures these forces are agitated. When Uranus stands in this aspect with other planets, thunderstorms are unleashed because the agitation reaches into the realm of electricity. The same aspect with Neptune changes the patterns of magnetism and leads to earthquakes. When Pluto is in quadrature primeval fire is activated on the Earth producing volcanic activity.

With *quintile* aspects (72°) there is a confrontation of adverse central forces with adverse universal forces. One is reminded of the Greek picture of the struggle of the Titans. This adverse quality works into the sheaths of the Earth and produces a variety of different kinds of stormy weather. Animals cannot protect themselves from such influences and easily become bad-tempered. Human beings are drawn into this if they do not compensate with heightened consciousness. People are often not in full control of machines if they are in this situation with a somewhat dreamy state.

We draw attention to these possibilities in the *Sowing and Planting Calendar.*

Oppositions

If planets form an opposition of 180° with the Sun or other planets then one can often discover greater intensification in the life processes of the plant. This begins a few days before, and increases up to, the point of the opposition. The forces of both planets penetrate, augment and intensify each other in the sphere of the Earth, supported by the zodiac working in the background. There are two groups of oppositions. Oppositions occur either in Light and Warmth constellations, or in Earth and Water constellations (Figure 7). Only rarely does an opposition not fall into this pattern. The advantages for plant growth are thus

Figure 7. Oppositions and conjunctions in the zodiac.

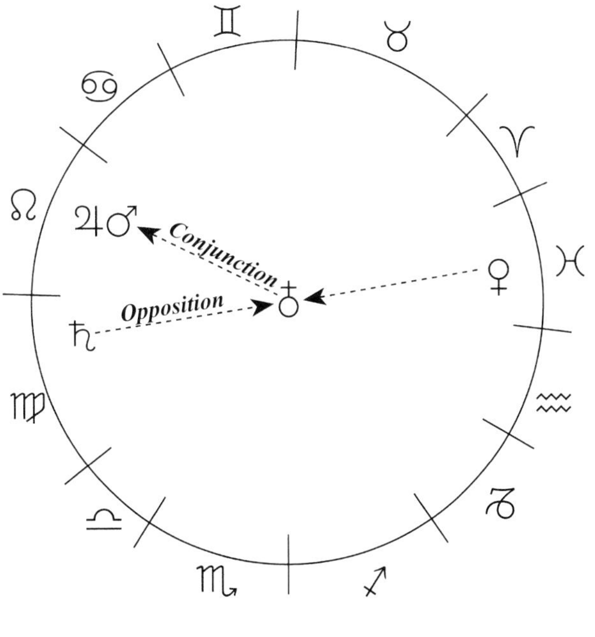

not always equal. In the first case the advantage comes with regard to flowers and seeds, while in the second case it appears in the leaf and root spheres. In this way the Moon's influence can at times be increased, but lessened at others.

Conjunctions

The effects of planetary conjunctions — that is when two or more planets stand together in the same direction — are quite different. Various kinds of plants were sown simultaneously on different experimental plots, and under different soil and climatic conditions, but all were planted when there were a number of conjunctions occurring at the same time. It was not possible to evaluate the results in the usual way as all of the plants died from fungal disease at the two or three leaf stage. This shows that in such a case the planets do not mutually enhance each other's influences, but seem to cancel each other out. If only two planets are in conjunction the effect is not so strong, but when several conjunctions occur together it is most marked.

Trines

When two planets are 120° apart (viewed from the Earth) they form a trine. In most cases we find that related forces are involved. The planets do not appear to have their own characteristic influences on plant growth and weather formation, but what can be observed is an enhanced effect of the constellations behind them.

The length of the influence which we attempted to find out through many hourly sowings, can last between five hours and three days, depending on the speed with which the planet concerned is moving on its course. In this way we can reckon with a one-sided impulse during a measurable length of time. This imposes itself on the sidereal Moon impulse, weakening it or strengthening it.

Summary

In his agricultural course Rudolf Steiner describes how the superior planets — Mars, Jupiter and Saturn — radiate below the ground, and how the siliceous rocks of the Earth reflect these forces back to the soil. He also describes how the forces of the inferior planets — Venus, Mercury as well as the Moon — are drawn into the soil through limestone in the Earth, and how clay has the ability to combine these two kinds of forces, thus making them accessible to the plant once the humus situation is right. If only half of these substances were present in the Earth we would attain grotesque plant forms. Clay is described as carrying Sun impulses within it, therefore earthly material and cosmic forces are brought into relationship with one another. Grotesque plant forms, on the other hand, come about at times of one-sided cosmic constellations.

This happens most of all when individual planets are prevented from exerting their normal influence due to eclipses and occultations. Then phenomena appear which lessen or even annul the plants' food-producing powers for man and animal, or adversely affect its reproductive forces.

We can therefore conclude that a harmonious collaboration of the heavenly bodies is required if the plant is to develop in a balanced way, both with regard to its nutritive quality and its reproductive forces. If disturbances occur we observe growth checks, fungal disease and pest attacks.

Weather Observations

The zodiac and the planets

We noticed quite early on in our experimental work with plants that it was necessary to take weather observations into account. It was soon found, for instance, that the sowing days for plants that produced good leaves always tended to be damp; in fact, they were mostly the wettest days of the month. Over the years, observation of the weather has led to the following results:

The general weather situation is connected with the rhythms of the planets and the zodiacal constellations that, at any given time, form their cosmic background. At the same time, therefore, a certain degree of zodiacal influence on the weather can be observed, as well as planetary effects. The third factor that comes into play is the four 'Elements,' which we will call Earth, Water, Air/Light and Warmth. These can be seen as also having a connection with the atmosphere that surrounds the Earth. Those who are in doubt should study air and space travel. Observations lead us to conclude that there is a certain fourfold pattern in the interplay of these various factors.

As regards the fixed stars, it is not so much a question of remaining unaffected, in cosmic space, but rather the effect on those that are continually disturbed by the planets. Once again we are concerned with that belt of fixed stars referred to as the zodiac. The influence which these constellations — or zodiacal regions — exert works through the elements of Earth, water, air and warmth, and in this way can be identified and observed. However, the planets also use these elements as a last resort in order to affect weather formation.

For the sake of clarity we will set out the various planets with their corresponding elements. The following order results:

Aries, Leo, Sagittarius	Warmth	Saturn, Mercury, Pluto
Taurus, Virgo, Capricorn	Earth	Sun, Earth, Ringall
Gemini, Libra, Aquarius	Air/Light	Jupiter, Venus, Uranus
Cancer, Scorpio, Pisces	Water	Mars, Moon, Neptune

Ringall

Since the mid-fifties, in connection with our plant and weather observations, we have repeatedly found that definite spells of cold weather occur whenever the Moon reaches a certain longitude in the zodiac. However, with squares (an aspect of 90°) and other aspects at this point then two or three cold days also follow. When we examined this more closely and asked friends in other areas for their observations we discovered that we were not dealing with a set point in time but a situation, which was gradually moving. Following this observation, which we have worked with for over two decades, more questions emerged. It has been suggested that there is a *twelfth* planet whose orbit is beyond Pluto's. We have called this object *Ringall.* Again and again we hear that scientists suspect the existence of another planet, but to date it has not been observed.

The planet's movement that we observed — based on the phenomena described above — amounts to 0.829° per year which would be an orbit of 399.6 years. Comparing this with other distant planets — which we know stand in front of 30° signs in seven-year rhythms — we find the following arrangement:

Uranus	84	(7 x 12 = 84 years)
Neptune	165	(14 x 12 = 168 years)
Pluto	248	(21 x 12 = 252 years)
Ringall	400	(33 x 12 = 396 years)

It appears that this is based on a higher order and so we venture to include this phenomenon in this little book.

There seem to be special laws governing the 'new' planets; at any rate, the effects of particular constellations show up additional factors that come into play here. We find Uranus is connected with electricity, Neptune with magnetism, Pluto with volcanic activity, and Ringall with cold-producing forces.

If one of the 'traditional' planets stands in front of a zodiacal constellation that has the same effect —with regard to the four Elements — as is characteristic of that particular planet, then that effect is intensified. However, if the planet passes a constellation that has a different effect, then its own effect is diminished, or even suppressed. For instance, if a warm planet, such as Mercury, is in Aries, then its influence is strengthened, but if, on the other hand, it is in

Taurus, then its warmth effect is not noticeable. Again, if it moves into a Water constellation, such as Cancer, then its warmth effect produces a tendency to rainfall. Let us take another example. When Venus is in a Light/Air constellation, we have blue sky and sunshine and a very clear atmosphere; if Venus moves into an Earth constellation, then the effect can be very similar, except that there is a greater danger of night frost. If it stands in front of a Water constellation, then we notice hardly any effect at all. There are similar examples that can be quoted with regard to the other planets in relation to their background constellations at any given time. For instance, if one of the planets which itself works through the watery element is standing in one of the Water constellations of the zodiac, then rain is expected.

Another factor that also influences the weather is to be found in the relationship of the planets with the Sun. Whenever the planets are in retrograde motion, so-called *loops* are formed. A planet's own characteristic effect can be most clearly identified during this time. For Venus and Mercury these times occur during their inferior conjunctions with the Sun, when these planets are moving in their orbits between the Sun and the Earth. In the case of all planets, whose orbit lies further out from the Earth than that of the Sun, both retrograde motion and loops occur in opposition. In both cases the planets are near to the Earth.

The overall weather condition is brought about by the rhythms we have described so far. However, the Earth itself affects it still further by the climatic zones surrounding the Earth.

The zodiac and the Moon

The change in the microclimate is also determined by sidereal lunar impulses. Whenever the Moon moves into another constellation a new aspect of the four elements comes into play, which despite generally uniform weather conditions brings subtle differences that can easily be observed. They are seen in slight changes of temperature, differences of humidity, wind direction and the formation of mist or fog, etc. These weather patterns usually last for two or three days and then new changes begin to be felt. These are understood as a result of the sidereal Moon in connection with the zodiac.

Oppositions

Besides the rhythms already described there are other aspects of the planets. From our vantage point on Earth, the angles that planets form between each other express in different qualities. For instance there are *oppositions*. In the case of Sun and Moon this is called Full Moon. Whenever two planets stand opposite to each other (at 180°) then the Earth stands between them. The forces of these two planets thus penetrate into the Earth sphere. A type of cosmic tension is created, and the weather picture is characterized by a high.

Conjunctions

If, on the other hand, two or more planets are close together — that is, in conjunction — then we can expect a low. When opposition and conjunction occur close together, then areas of high and low pressure are often very close together. When a high occurs factors of a more cosmic origin express themselves in warmth and cold; in a low earthly influences, working through water and air, predominate — that is, mist, rain, showers and storms.

Trines

Another characteristic angle between the positions of the planets which also play a strong part in weather formation, is the trines that arise when two or more planets stand 120° apart. Its effect is always positive on the element of the zodiacal constellation which forms their background. These positions almost always fall into the same realm of forces — that is, they are subject either to a warmth impulse or a watery one, an influence of light, or an earthly one, and the day in question will show the corresponding tendency in the weather.

Sextiles

With a sextile, a 60° angle, on the other hand, it has been confirmed repeatedly that there is a connection to the watery element and, depending on the planets involved and their zodiacal background, this aspect manifests itself either as mist, rain or showers. At any rate, there is always the tendency to precipitation.

Quadratures

The situation is different when the planets stand at right angles to one another. For Sun and Moon, this aspect occurs at the quarter phases of the waning or waxing Moon (Half Moon). With other planets, these particular positions have the strongest influence when one, or even both, are *new* planets — that is, Uranus, Neptune, Pluto (or Ringall). Aspects of 90° involving *traditional* planets are rarely factors in weather formation. However, with Uranus at 90° to another planet, we can expect thundery conditions, often accompanied by storms and showers. Neptune in such a position mostly gives rise to earthquakes, or in earthquake-free regions, to natural disasters caused by severe storms and violent rain. Pluto in that position causes volcanic eruptions, and in regions with no volcanic activity there will be violent storms.

This latter phenomenon must be expected when Pluto and its corresponding planet are parallel to each other at the same time. Aspects of 45° and 135° show the same tendencies but with less intensity. With the same Ringall aspects we get one-sided cold weather with storms, often in regions where such cold spells are unusual.

Quintiles

The most severe natural catastrophes are produced when the new planets are in what can be called a *pentagonal* aspect with other planets — that is, at an angle of 72°. When planets stand in this aspect to each another there is a higher incidence of accidents involving not only aircraft, ships and motor vehicles but also other machines and animals. On such days if one tries to observe how one reacts and how others react then it becomes clear that one's consciousness is not sufficiently awake to what comes from the outer world, but is more introspective. These indications in the *Sowing and Planting Calendar* should really be taken as a warning to activate one's wakefulness on such days. Since we started giving this indication we have received innumerable letters from people who were involved in accidents on such days and who learnt to be more attentive. To a lesser extent we find similar effects during aspects of 144° and 36°.

Whenever we have more than five or six millimetres of rain, quadratures, quintiles or planetary aspects related to them are often there as causal factors.

Summary

In more than forty-five years of weather observations in connection with plant growth experiments we have not been able to study all possible aspects since these depend on the length of orbit. The distant planets take a long time to orbit the zodiac once — Saturn takes 30 years, Uranus 84 years, Neptune 164 years, Pluto 248 years (and Ringall approximately 399 years). The times of the nodes — especially the nodes on the ecliptic when the planets are at the points of orbital intersection — are reflected in weather formations. In the case of Mercury we notice its effects over the course of one day; for Venus and Mars over two days; and for planets with lengthy orbits their influence can stretch over many months, registering as extremely mild winters or very cold summers. If quadratures and quintiles occur at the same time as the planets are in nodal positions we often experience particularly severe natural disasters. During longer nodal periods animal pests are more prevalent. In the case of Saturn these animal pests affect the roots; in that of Jupiter the leafy region of the plant is most affected.

While we must always be prepared for new and unexpected phenomena, especially where longer planetary orbits are involved, it can, nevertheless, be said that weather forecasts made on the basis outlined above have so far proved to be 75–80 per cent correct.

These notes are intended primarily to stimulate the reader to make their own observations. It is advisable to start with simpler rhythms, perhaps with the Moon's sidereal rhythm — that is, the rhythm that forms the basis of the *Sowing and Planting Calendar* — and its effects on the local climate. Brief notes of special events should be made and then perused and related to the astronomical events of the days concerned. It is surprising to note that after only a few years some degree of certainty of judgement can be achieved, although it is always a good idea to assist one's memory with short notes made at the time.

Sand Forms

When staying by the sea we walked on the beach and saw how the water had inscribed delicate forms in the sand. We hardly took any notice of this but after a while we could see that there were quite different forms, which were repeated. We began to take note and compared the transitions to new forms with the calendar. We discovered that whenever the Moon moved into a new constellation the water left behind different forms and patterns in the sand. These appeared when it had been high tide, the last wave had crashed down, and the water was ebbing away. As the water retreated different organic forms were left in the sand, mainly plant-like forms.

Only with the Moon in Sagittarius did one see animal-like forms. When the Moon was close to the Earth (Perigee) hard grooves appeared but without any real pattern. In Figures 8–15 we can see such forms. This shows the influence of cosmic rhythms not only in living organisms but as expressed directly in the elements.

Figure 8 (left). Sand forms when the Moon is in Cancer.
Figure 9 (right). Sand forms with the Moon at perigee.

Figure 10. Sand forms when the Moon is in Virgo.

Figure 11. Sand forms when the Moon is in Capricorn.

Figure 12. Sand forms when the Moon is in Aquarius.

Figure 13. Sand forms when the Moon is in Leo.

Figure 14. Sand forms when the Moon is in Sagittarius.

Figure 15. Sand forms when the Moon is in Aries.

Halley's Comet

Halley's Comet appeared in 1910/11. From descriptions of old farmers we know that the year 1910 brought some catastrophic weather in Europe, and in 1911 there was a severe drought. which left farmers without fodder for their animals. None could be bought nor could the animals be sold since butchers had a surplus. The animals grew thinner and were driven into the forests, where they led a meagre existence, feeding on twigs, bark and moss. This drought came to an end with storms, which lasted for a whole week and were accompanied by deluges, which caused great damage. In our local Dexbach area (where we have been carrying out our experiments), all the topsoil was swept off the slopes and the floods left gullies throughout the countryside. Just below one of our fields is a broad ditch, 10 metres deep and about 15 metres wide, which appeared at that time. It was a period of natural disasters that left a deep imprint in human hearts.

Halley's Comet reappeared in 1983 and was closest to the Sun and Earth in 1986. There were similar extremes in 1984. Periods of drought alternated with floods, accompanied by huge hailstorms — the worst in living memory.

While Halley's Comet was moving through Gemini it blotted out the influences, which the inner planets Moon, Mercury and Venus always transmit from this region. As far as weather and plant growth are concerned it appeared as if these three planets did not exist any more. A drought followed that lasted for weeks.

At the beginning of September 1984 the Sun was at a 60° angle with the Comet. Venus was square to it, Mars and Uranus at an angle of 150°, and it began to rain. Three weeks later as much rain had fallen as one would expect during an average year. Due to the late sowing the summer grain did not ripen in the more mountainous areas; it germinated in the ear and could at best only be used as animal fodder.

When in February 1986 Halley's comet came near the Sun and in April 1986 near the Earth, further disturbances of lunar and planetary rhythms occurred.

The Stages of Development of our Planetary System

From Rudolf Steiner's descriptions of evolution we know how our solar and planetary system has developed from a subtle warmth element through various stages of densification to our own earthly conditions. There were four great stages. The four traditional elements result from the epochs of creation. They are the gifts of the gods, which have streamed from the impulses of the heavenly zodiac. During the four great stages the planets emerged. There are therefore connections between the planets and the zodiac, as well as the four elements. Each of the planets is able to mediate a particular element and thus convey starry forces to the Earth. One can repeatedly find the following relationships:

The constellation	**The planets**	**The elements**
♈ Aries	♄ Saturn	
♌ Leo	☿ Mercury	Warmth
♐ Sagittarius	PL Pluto	
♒ Aquarius	♃ Jupiter	
♊ Gemini	♀ Venus	Air/Light
♎ Libra	⛢ Uranus	
♓ Pisces	♂ Mars	
♋ Cancer	☾ (Moon)	Water
♏ Scorpio	♆ Neptune	
♉ Taurus	♁ Earth	
♍ Virgo	☉ Sun	Earth
♑ Capricorn	Rg (Ringall)	

The seven main planets predominantly use these four elements in order to influence the weather, as well as plant growth. Planets beyond the orbit of Saturn use central forces mainly to release their own impulses — Uranus, electricity; Neptune, magnetism; Pluto, volcanic activity; Ringall, cold. In all four cases we are dealing with hardened, or *dead*, forces of the four traditional elements.

The Moon occupies rather a special place in this. While the planets are able to transmit mainly one element, the Moon has the capacity of transmitting all four, although by nature its own influence is watery. Rudolf Steiner describes one ability of the Moon which puts it ahead of the planets of the fourth stage of evolution, namely that of growing old. This is something, which really belongs to a future fifth stage.

It was clear from our experiments that the Moon can certainly work through water and mineral salts in its regular watery way, but once it finds enlivened Earth then another way of working becomes apparent, which points towards a future development. It then transmits all four traditional elements, including the forces which come from the zodiac. We can speak of zodiacal forces, that the Moon transmits to the plant via the elements, whenever it finds enlivened soils that have been activated by compost. These are soils where neither mineral salts nor animal substances lead a life of their own. The bee living in the social structure of the hive is also orientated to the future, and likewise it lives with this future aspect of the Moon. With Venus the bee lives only through the light, and with Mercury only through warmth. Other animals are much more one-sidedly orientated, both with respect to the Moon and with the other planets. Only mammals, which produce milk, have some connection with the Moon's transmission of zodiac forces. For this reason milk seems to be a unique form of nourishment.

If we look at comets we are dealing with astronomical phenomena that do not fit into the rhythms of our present fourth stage of evolution. They appear and disappear again. They claim a certain 'freedom' that neither the Moon nor the planets have. They have remained at the third stage of evolution and when they appear they bring disturbance to the other planetary rhythms. Above all the Moon's special degree of evolution appears to be an irritation to Halley's Comet because it attempts very hard to bring everything into confusion. One is reminded slightly of children; one child has built a house with blocks and another child, who can't yet do that, wilfully knocks it down. The comet is assumed by scientific research to consist most likely of a conglomeration of meteorites and ice particles, and the tail is thought to be made up of carbonic oxide, cyanide and hydrogen. It appears that the planet Ringall, which closes the outer ring of the cosmos, takes hold of certain substances by means of its forces of cold, and when the comet comes into a solar system it forms ice in order to transmit for a while an extreme condition of Earth to the comets, which have not yet reached the fourth stage of earthly evolution.

Planting Time

As we described on pages 32–40, the Moon's revolution round the Earth takes twenty-seven days. When it has reached the lowest point on its path in Sagittarius it begins to ascend. It moves across the sky in arcs that become higher every day, quite independent of its phases. When after fourteen days it has reached the highest point of its path in the region of Gemini it begins to descend. During the period of descent the forces and flow of sap move more into the lower part of the plant, and the plant is more strongly orientated towards the Earth. If one transplants it during this time new roots are quickly formed and find a new connection with the Earth. This tendency is apparent with transplanting but not with sowing. We recommend this time as planting time. At the same time it is suitable for manuring meadows and fields and working in green manure. Cuttings

Figure 16. Carrots sown on the same day and transplanted under different conditions, from left to right: sprayed with nettle manure (Br), sown on root days and not transplanted (EK), no Preparations sprayed after transplantation (O), Preparation 500 sprayed after transplantation and 501 later on.

when taken during this period are quick to root. At the appropriate time of the year planting time is also suitable for felling trees for timber. Root crops harvested at this time store better.

When transplanting one can also use other rhythms, which work in collaboration. This will enhance the effects on the various plants. When transplanting tomatoes one can choose days which encourage that fruit type during planting time. These are Leo days. One may choose Leaf days when transplanting lettuce or cabbages — Cancer and Scorpio days during planting time. For the flowering plants the days in Gemini and Libra can be used. We can also point to another good possibility, namely to buy cut roses in a shop as long as they have not been over-cultivated. When they are fading cuttings from them can be put in pots on Gemini days. This encourages rooting and blossoming so that one can always have roses in bloom in the house.

Since the upward movement of plant sap is weaker during planting time we also find that this is the suitable time for pruning trees and cutting hedges.

Figure 17. Rose-cuttings should be planted when the Moon is in Gemini and kept moist for the first days.

Good Friday and Easter Saturday as Sowing Days

Even from our experiments carried out over thirty years ago and repeated many times since then, we discovered that these two days could not be recommended either for sowing or planting. Sowings made on these days had a bad germination rate, poor growth in the early stages and produced small yields. Seedlings transplanted on these days did not develop good roots and most of them did not thrive.

This negative influence on plants begins in the early part of Good Friday and ends on Easter morning around sunrise. For this reason we have designated these days as 'unfavourable' in *The Sowing and Planting Calendar*. During the last few years there have often been planetary oppositions, which brought good weather, and at the same time the soil was in good condition. After Easter the weather was sometimes no longer so good, and we had complaints from those active on the land that they had not used the lovely days.

In 1984 there was a planetary opposition on Good Friday and a Warmth trine on Easter Saturday. The soil was in the best possible condition and the weather was perfect for sowing grain. So we decided to sow three different grains, summer rye, summer wheat and a variety of barley. The available area comprised about 1000 square metres and had received a light dressing of compost in the autumn. The sowings were made uniformly and later divided into strips for comparisons of hoeing and spraying. In case the day of sowing proved unsatisfactory, we wanted to discover how we could help plant growth using favourable cosmic influences for hoeing, and using the preparations. We had five variations:

1) 0 — control plot
2) AK — hoed on Leaf days and sprayed with 501
3) WK — hoed on Fruit days and sprayed with 501
4) EK — hoed on Root days and sprayed with 501
5) LK — hoed on Flower days and sprayed with 501

All this was repeated three times. Around the time of sowing we applied Preparation 500 (Horn Manure) three times. The Moon was in Sagittarius, which was a Fruit day. The seeds germinated quickly and so we assumed that it was different in this year. However, the plants soon began to look sickly and they did not really get beyond their early growth and grew slowly. It seemed as if there would be no real ears. At last some quite meagre ears appeared. In the meantime we had been hoeing and spraying regularly but the whole area made one almost want to cry. When harvest time approached the grain did not ripen. The corn did not harden properly.

Then it began to rain. After two weeks there was now and again a clearer day. The grain was no riper, but the Leaf day variations had already begun to germinate on the stalk, so we set about threshing and drying it. Even when we were carrying the sacks we noticed that their weights varied. When they were dry they showed the following results in tonnes per hectare:

Varieties	1) Control group	2) AK	3) WK	4) EK	5) LK
Barley	1.4	1.86	1.97	1.77	1.84
Summer rye	3.3	2.98	4.20	3.65	3.60
Summer wheat	2.2	2.54	3.80	2.80	3.00

The control group was hoed on Fruit days but was not treated with Horn Silica (501) three times. The yields were low. In the case of the rye and the wheat the effect of the Silica spray on Leaf days had gone into the straw and brought an increase in yield while with the barley, there was a noticeable increase. All three cereals responded well to cultivations on Fruit days. It therefore becomes clear that the inferior quality of a sowing day can be compensated for quite considerably by cultivating on favourable days. However, Good Friday and Easter Saturday will continue to be marked as unsuitable in our Sowing and Planting Calendar in the future. The event that happened almost 2000 years ago between the Sun and the Earth leaves its traces in the Earth every Easter. There may well be a reader who wishes to write a philosophical or a theological book about it.

Figure 18. Dr Storch with tomato plants.

The Influence of the Constellations on Bees

The bee with its many tasks occupies an important place in life on the land. People understandably think that the beehive is a little honey factory. However, the experts know that the preparation of honey is only one of the bee's beneficial activities. There are others, which are more important for the whole of nature and plant growth than the honey yield.

It is well known to the plant grower and the plant breeder that for many plants pollination takes place through the honeybee.

Bees also provide beeswax, bee venom and propolis, which are used for medical purposes. At Christmas when a real beeswax candle is burning we can experience with great wonder how light forces were brought into the wax by the bee and how these are now liberated in the flame.

In his agricultural course Rudolf Steiner highlights a result from his spiritual scientific research when speaking of how certain forces, working from the periphery, are so distributed among the world of insects that new plant growth is possible again and again. It is therefore necessary for those working on the land to see that there are bees around. If beneficial insects are absent then a certain one-sidedness makes itself felt and this results in insect pests on a large scale.

In his lectures about the real nature of bees Rudolf Steiner makes this even clearer by adding one other fact. Whenever an insect visits a flower it secretes minute amounts of its poison over the flower. 'If this did not happen ... the necessary poisons would not reach these plants and the plants would die out after a while.'

We can therefore see that just as we need the cow to maintain the fertility of our field by giving the right conditions for plant growth in the soil through its manure, the bee works at the opposite pole on the upper part of the plant, so that new plant growth can take place. It is true that these are very subtle interrelationships, which cannot be easily observed.

Recognizing the importance of beekeeping in agriculture we decided to take it up ourselves. In this way we were able to study the bees' activity much more closely, and we soon realized the importance of observing the bees once a day, sometimes more often.

As a result of plant experiments and weather observations over many years we had become aware of the great variety of cosmic rhythms. Suddenly a question now emerged. Is the life of bees subject to similar, or the same, individual cosmic rhythms as we had observed in plant life and weather formation? We therefore put some bees on the balcony of our house for two years so that we could observe their activity at any time of the day.

After a short while we noticed the bees' daily behaviour was in no way uniform. There were days when the bees returned laden with pollen. When we tested the honey sac with a needle we found it was empty. There were other days when the bees were eagerly gathering nectar but despite the sunny warm weather they were not bringing in any pollen. On other days the bees did not fly much and they brought neither pollen nor nectar back home. Then there were suddenly times when pollen and honey gathering stopped for two or even more weeks. However now and then there were days when the otherwise peaceful bees shot through the air like lightning and anyone who got in their flight path would certainly be stung.

After we had observed these different phases for two years we had formed a picture of how cosmic influences worked in certain patterns. For example, the times of good pollen gathering coincided with Venus standing in the constellation of Aquarius, Gemini, or Libra. A longer period of making honey coincided with Mercury standing in the constellations Aries or Leo. Those are just a few examples. If the beekeeper can calculate when he can reckon with the first gathering of pollen in spring he will adjust his feeding so that the first brood is just enough developed for the pollen to be used properly.

Again and again there had also been groups of days (apart from the larger rhythms mentioned above), which produced a good honey flow or a lot of pollen, and we were able to find a connection with the sidereal Moon rhythm. As the Moon moved through different regions of the zodiac the bees gathered nectar, or at other times pollen, and sometimes they did neither. One was reminded of a passage in the lectures 'The Nature of the Bees' where Rudolf Steiner answered one of the beekeepers who had asked the following question:

> With regard to the influences of the zodiacal constellations on honey production, the peasants lay great stress on sowing seed when the Moon is in Gemini and so on. The question is whether

> this idea as to the constellations of the zodiac is founded on external data, or if there is more than this in it?
>
> *Dr Steiner*: You see, gentlemen, today these things are never dealt with scientifically. But one can treat them scientifically. On the whole colony of bees, as such, there is as I told you, an influence. The bee, and above all the Queen is, in a certain sense, a Sun creature, and thus all that the Sun experiences in that it passes through the zodiac, has the greatest influence. But the bees naturally depend on the plants, and here indeed, the sowing, the scattering of the seed can be very much affected by the passage of the Moon through a zodiacal constellation; this concerns the preparatory substances the bees are able to find in the plants.
>
> These things are by no means fanciful, but as a rule they are represented quite superficially; they should be much more deeply studied.

So the real question is what the bees find as substance in the plants. Does the plant only release nectar or pollen under certain influences, which are determined by the environment?

More and more new questions arose. It also became clear that bee colonies reacted differently to the beekeeper's handling on different days. On some days they went on working without disturbance while on other days a veritable chaos in the colonies followed the handling which developed into an orgy of stinging.

In the method, which avoids letting the bees swarm, the colonies are inspected every nine days. This links with the development of the queen who needs nine days from egg laying to the sealing of the cell. Depending on the day when one begins the first handling in the early part of the year one becomes involved in rhythms, which were shown to be fourfold in our plant experiments. They depend on the sidereal Moon rhythms and have been explained in the earlier chapters of this book in connection with plant growth and weather formation.

We were able to find out in our experiments while handling bees, that the bee experiences its activity in a certain relationship to the course of the Moon. When the Moon passed through the constellations of Aries, Leo and Sagittarius the bees brought much more nectar, while the Moon's passage through the regions of Gemini, Libra, and Aquarius brought more concentration on gathering pollen.

Figure 19. Apiary for tests at Treisbach.

Figure 20. Beekeeping at Dexbach.

When the Moon was in Taurus, Virgo or Capricorn the bees were busier building in the hive.

The bees live with the surroundings as with a cosmic clock. The plant and the weather operate with these laws too, and for this reason there is a wonderful harmony between the plant and the bee:

Plant	Bee	Weather	Elements
Fruit/seed	Nectar gathering	warm	Fire/Warmth
Flower	Gathering pollen	sunny/light	Air/Light
Leaf	Honey processing	moist/watery	Water
Root	Comb building	cool/cold	Earth

Light and warmth are the priorities for breeding queens and building up the colony. When harvesting honey there are marked differences in crystallization. A comparison of these phenomena has not yet been completed.

PART TWO

Soil, Manuring and Composting

The Soil

The basic mineral substance of the soil that we cultivate derives from the weathering of rocks and stones. If this has resulted in one-sided conditions, then these have to be balanced by certain mineral additives. This is not then a question of manuring, but of harmonizing the basic mineral components. Through the process of weathering the formation of secondary clay minerals is always made possible. Powdered basalt is recommended for this purpose, and is added in small quantities to the manure or compost as the heaps are built up. The process of weathering or decomposition releases dormant forces in the rock, which strengthen the life processes in the soil, and make for a better co-ordination of mineral and organic substances.

The real secret of enlivened soil lies in the right sort of decomposition, which is activated by the soil organisms. The soil then has the capacity to transmit cosmic forces. These forces need an enlivened-Earth element as the basis of their influence. Rudolf Steiner points this out in his course on agriculture, when he says: 'Manuring means enlivening the soil ... Life must be brought close to the soil, the Earth itself.'

Experiments with plants

From almost forty-five years' experience it has repeatedly become clear that the condition of the soil is a decisive factor in determining whether the differences in a plant's rhythms can be seen or not. Animal substance, which has not been composted, and high doses of mineral fertilizers produce the greatest hindrances. Even raw humus that has not been broken down in the soil can interfere with the working of cosmic rhythms. A good amount of well-rotted humus of 2–2½% or more, gives a good guarantee that rhythmic differences will appear in accordance with the constellations. It is also better to avoid artificial watering.

Experiments with potatoes and cucumbers

In experiments with the constellations we have attempted to discover if we get the same results from our sowings despite environmental pollution and confusion caused by satellites. We can, therefore, show yields of the potato *Grata*, in both cases cultivated from commercially-grown seed. They were planted from 1965 seed on very good soil, and from 1982 seed on poor soil. We have added the yields, which Mr Schwarz achieved on the Troxlerhof in 1968. The overall pattern remained the same although one or other planting was favoured or hindered by weather conditions and planetary constellations.

The experiments with cucumbers (Figure 22) show quite clearly that the yields on heavier soil were higher for normal annual rainfall but the characteristic tendencies remained the same, with good yields deriving from Fruit days.

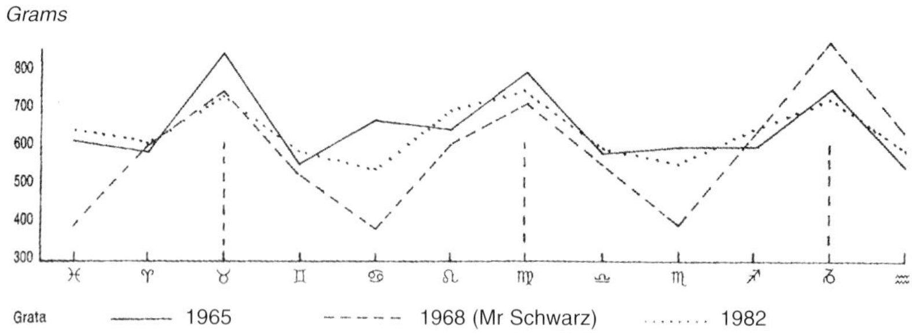

Figure 21. Grata yield.

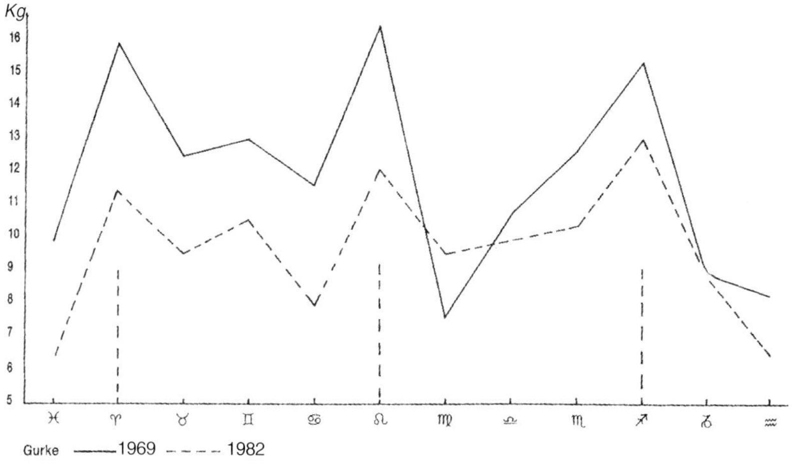

Figure 22. Cucumber yield.

Trials with dwarf beans and soya beans

Here we have the same problem as before. Figure 23 shows yields from the green harvest of dwarf beans using twenty plants. The comparisons are based on sowings made in 1965 and 1982, on different soils and with different climatic conditions. Figure 24 shows yields from soya and dwarf bean seeds in 1982, using one hundred plants. In the first case we are dealing with fruit yields, and in the second case with seed yields. The series of experiments shows the same inner dynamic, which is produced by the cosmic rhythms we have been studying for thirty years.

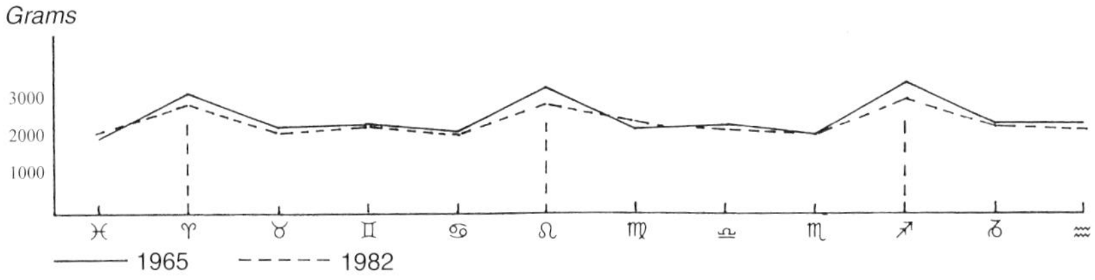

Figure 23. Dwarf bean yields.

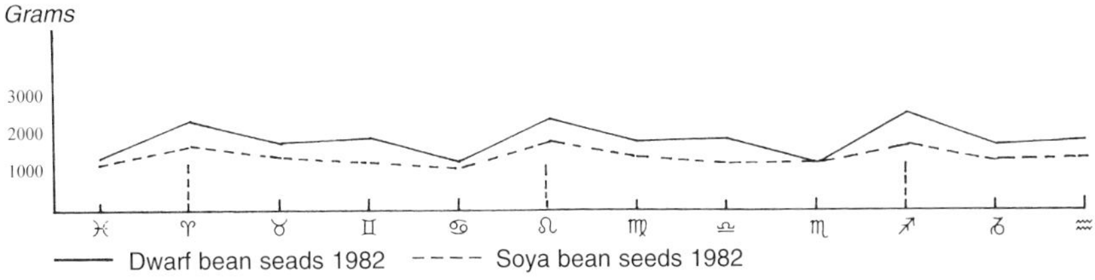

Figure 24. Dwarf bean and soya bean yields.

The Manure Process

In organic farming manure is usually applied directly to the soil. This means that either fresh manure, or manure in combination with a green manure is worked into the soil and left there to rot. With skill one may achieve good results with this. Farmyard manure can only be applied when there are no crops growing in the soil, and it is necessary to look after it in the meantime. It is sufficiently well known that there is much loss of valuable substances in the big heaps of manure, and for this reason there is a direct treatment of manure in biodynamic agriculture. In many cases this begins in the stable or cowshed. One can sprinkle small amounts of powdered basalt or add preparations to encourage the decomposition of the manure, or add a compost starter to guide the fresh manure in the right direction, so that these substance losses can be avoided as much as possible. After the shed or stable has been cleared the manure is placed in heaps and kept moist so that the bacterial and fungal organisms can soon begin their work. Once the heap has reached the required size preparations are added to it, as well as a final covering of peat. We experimented with manure from different animals and made heaps of comparable sizes, adding the compost preparations to some, but not to others. It became evident even after only a few days that there were marked differences. The treated heaps warmed up during the week up to 26–30°C, depending on the kind of manure and after about ten days they showed a good amount of bacterial activity on the surface. They also did not collapse but retained their shape. The heaps that had not been prepared heated up to 60–70°C according to the type of manure. They emitted quite a lot of steam and one could tell by the smell that some volatile substances were escaping and that the heaps were sinking quite visibly.

The remaining material was analysed and two months later new samples were taken for tests. It became clear that the prepared heaps were permeated with fungi showing that favourable processes of decomposition were under way, whereas the unprepared heaps showed dry mould. We know that it takes many months for this to rot properly when it is worked into the soil. Tests demonstrated that the treated

Figure 25. Compost trials.

Figure 26. Manure trials, potatoes and wheat.

heaps had a much better balance of materials. Two months later further samples were taken. In the meantime the treated heaps were found to be permeated with red compost worms. The untreated heaps only achieved a similar condition at the end of one year. The high temperatures at the beginning delayed the process of decomposition quite markedly, and encouraged substance loss.

In the years following these experiments different types of manure with different treatments were compared, and then the qualities of the various sorts were tested on different kinds of plants.

At the same time we also carried out experiments with various organic manures, which are available to the trade. Even here there were great differences depending on whether the manure had been composted or worked into the soil while fresh. When we carried out quality tests on the plants to determine their ability to store well and their power of reproduction, we found that it was better to pass the manure through the decomposition process. It should also be noted that the four elements — Earth, water, air and warmth — should work in harmony so that combustion, putrefaction or peat formation do not occur, since these one-sided conditions impede the activity of fungi, bacteria, worms and other organisms in the decomposition process. All the material should move towards humification so that we can add this end product to the humus in the soil. Un-rotted dung hinders the influence of cosmic forces in the soil and diminishes the effects of biodynamic sprays.

Green manuring

Since there is never sufficient manure from the farm itself to cover the organic material requirements of farm, garden and orchards, we can grow certain catch-crops for the purpose of turning the plant into the soil as green manure. In this way soil organisms, such as bacteria, worms etc. are provided with sufficient nourishment to ensure that their propagation and fungal activity is also stimulated. In the excrement of these soil organisms we find converted organic substances and certain newly-isolated minerals, so that by this means we are able to achieve part of the manuring programme. These organisms also bring about both the aeration and organic transformation of the soil.

The question of when to sow green manure crops depends on the use to which they are to be put. If foliage is meant for animal fodder, then you sow on a Leaf Day. If there is no need for that, then particularly legumes should be sown on a Root Day, because they will then develop nodules more profusely, and thus increase the nitrogen in the soil to an even greater extent. The best time for turning the green manure into the ground is during the descending Moon that is also the planting time, because transplanting is also more successful during this period. This is also the right time for spreading compost and liquid manure. If, for instance, you spread manure on fields and pastures during the ascending Moon, it tends to be carried up by the growing plant and left hanging on the grass. If, however, you do your spreading during the descending Moon, the manure is carried down to the soil by the descending forces and drawn into it by earthworms and other organisms.

The changes that take place when green manure is worked in can be increased and speeded up if one sprays the manure concentrate as well. It is a preparation developed with the aid of biodynamic compost preparations at our experimental station. We questioned how it might be possible to introduce the positive influence of compost preparations into the soil, more often than was possible within the limits of crop rotation and spreading animal manure. It also seemed desirable in connection with Demeter quality that one gave stimuli via the preparations more frequently, when a person was converting to biodynamics.

Manuring

Manure, which is meant to provide an enlivening element for the soil, should be spread out over wide areas in the autumn so that it can be broken down in the soil organism in winter. Special manures which are adapted to the needs of individual plants and which are applied in seed drills or in planting holes should be specially prepared. If one has not put substances based on animal products — such as hoof and horn meal, bone meal, bristles, wool, feathers and even guano — through the composting process and would like to use them for special purposes, then one can get them decomposing quickly by a special method. Use mature plant compost, taking 80% compost and 20% of those other substances. Make small heaps, ensuring they are moist and cover them with straw, or failing that, with a tarpaulin or cover. Mostly one has to apply more water. After about five to six weeks a white layer appears on the small heaps showing that the first phase of decomposition has taken place. If one uses this special manure, now the danger of a fungal attack is much diminished compared to the fresh application of the animal substances mentioned above. Nevertheless the ideal remains the total transformation inside the compost heap.

Manure Concentrate Developed by Maria Thun

Over many years biodynamic compost preparations have already proved to have a positive effect both in assisting the rotting down process of compost and manure, and in ameliorating the smell. They also aid soil formation and improve soil structure. Naturally, the question arose as to how to place this beneficial effect into the soil more often than is possible through ordinary manuring and composting, over the course of a crop rotation. It seemed desirable, too, for people in the process of changing over to biodynamic methods, to be able to use the compost preparations more frequently.

The manure concentrate is an intensification of the influence of the compost preparations, including the spray Preparation 500.

Other questions had arisen earlier. About twenty-five years ago investigations at a research station in Freiburg showed that plants grown on limestone soils contained far fewer residues of a certain type of radioactive fallout, than plants of the same kind that had been grown on siliceous soils. Experiments carried out in this connection over a period of eight years, showed clearly that the shells of chicken eggs have a significant role to play with regard to the calcium process in the soil, and in regulating its pH value.

The usefulness of basalt was investigated, too. It was found that on one hand, basalt can be added to compost and manure in the form of grit; it then supports the continued rotting processes in the soil, which in turn favour the formation of secondary clay-minerals. On the other hand, if added in a finely powdered state, basalt acts in a nitrogen-fixing capacity. Experiments involving the use of Horn Manure (500) and Horn Silica (501) were carried out and the results for both soil and plant led to the idea that basalt as well as eggshells should, in homeopathic form, be incorporated in biodynamic work.

Cow manure was chosen as a medium through which to combine these three aspects. Cows were fed sufficient roughage for the dung to be fairly firm. To five buckets of cow dung (without any admixture of straw) we add 100 grams (about $3^1/_2$ ounces) of dry, very finely crushed eggshells and 500 grams (about 18 ounces) of basalt powder.

All this is put into a wooden barrel. Then it is thoroughly mixed with a spade for a whole hour so that it becomes one dynamic whole. To start off a kind of composting process we put one half of the mixture into a barrel from which we have first removed the bottom and which we have buried 40–50 cm (about 16–18 inches) into the ground; the remainder of the soil is thrown up against it on all sides. Into this first half we insert one portion of each of the compost preparations but separately. The second half of the mixture is then added and is treated in exactly the same way. Then for 10 minutes we stir 5 drops of Valerian preparation in 1 litre ($^1/_4$ gallon) of water, and pour on. Finally, we cover the barrel with a wooden lid or board and leave it. It should stand outdoors, of course. After about four weeks the whole mass is 'dug over' once more and then left for another two weeks, after which it is ready for use.

When carrying out tests with Horn Manure (prep. 500) we take 30 grams (about 1 ounce) as one portion and use it, stirred in 10 litres (2 $^1/_4$ gallons) of water, for 14 hectares (i.e. just over half an acre). In the case of the manure concentrate we use 60 grams (approx. 2 ounces) for the same area and in the same amount of water.

The best effect on the soil processes is achieved after three sprayings. These can be taken from the same mixture over a period of one or two days. Whereas Horn Manure (prep. 500) and Horn Silica (prep. 501) have to be stirred for one hour to attain their full efficiency, we need only about one-third of the time with this concentrate, i.e. 15 to 20 minutes. The special preparation process described above makes for a labour-saving procedure when it comes to using it.

This manure concentrate is, however, no substitute for Horn Manure (prep. 500) that is used at the sowing times and has a direct effect on the plant. The former stimulates the soil metabolism by activating the micro-organisms and thus results in better decomposition of organic matter and better soil structure. It is recommended, therefore, to spray it when green manure or rotted farmyard manure is spread, and also on autumn-ploughed land. When sprayed on pastures after grazing, the soil metabolism is stimulated and growth enhanced. Control experiments showed considerable increase in yields.

This manure concentrate can be warmly recommended to the practical grower.

The development of Manure Concentrate Preparation

The Manure Concentrate Preparation was developed between 1958 and 1972 in numerous experiments and trials, and in 1972 we reported on it for the first time. Meanwhile it has been applied all over the world and appreciated in enthusiastic reports. Now I feel obliged to detail its history.

In the 1950s several nations had performed atmospheric atomic bomb tests leading to the pollution of many parts of the world with radioactive Strontium 90. Many research institutes in the USA, Britain and Germany have measured this. Plants of the same kind grown on different soil have been investigated at a research institute in Freiburg, Germany. Plants grown on silica-rich soil contained high residues of Strontium 90, while plants grown in the Rhein valley contained less. The same kind of plants grown on lime soil contained only traces of radioactive Strontium.

Investigations of animals revealed a high residue of radioactive Strontium 90 in the bones of the skull. To the horror of medical practitioners investigations of human beings revealed that breast milk had the highest concentration of radioactive Strontium 90 of all human substances. It has since been discovered that organisms with an unbalanced lime metabolism erroneously incorporate more Strontium 90, and have a higher risk of contracting leukaemia.

In the USA Dr Ehrenfried Pfeiffer performed similar investigations and obtained comparable results. After discussions with him we planned to perform joint experiments, he in Spring Valley (New York) and we in Marburg (Germany). We decided to grow plants on organic lime soil and study the influence of Strontium 90 incorporation. In Marburg we chose the particularly vulnerable plants oats, celery and tomato. The trials were conducted on our test field 'Gartenweg' which had sandy soil and had been cultivated biodynamically for six years. Dr Pfeiffer was also concerned about a degradation of the biodynamic preparations.

We decided to work with the following substances: eggshells of chicken and duck, snail shells, bone meal, oak bark, lime of algae, wood-ashes, limestone and ground basalt. The substances were ground to a grain size of 0.2 to 0.5 mm and a certain amount was floury. The substances were spread using the seed drills prior to sowing, and after sowing were covered with soil. The plants were then cultivated in accordance with their type (root, leaf, flower, fruit).

The total crop of oat — in the state of being 'milk ripe' (milchreif) — was analysed, but only the leaves and crops of celery and the leaves and fruits of tomato were tested. An unambiguous result was obtained: plants grown with eggshells and ground basalt did not incorporate or store any radioactive Strontium 90. It was a miracle!

So we thought a lot about a way to bring these two substances to the biodynamic farms and we prepared a fermented liquid — 'liquid manure' — from eggshells, ground basalt and water. One part was sprayed on the soil while the other one was mixed with horn manure preparation, and then sprayed on the soil. For comparison one plot was sprayed with pure water. Five different plant species were grown. The results obtained were clearer for the 'liquid manure' than for the mixture.

We also looked for other means to bring the effect of these two substances to the farms. We put eggshells into cow horns and also ground basalt into other cow horns and placed them all into the Earth, one during the winter and the other during the summer following the procedure for the biodynamic preparations. New trials were made: each preparation was stirred for one hour and then sprayed partly like horn manure on the soil and partly like horn silica during green plant growth.

The growing plants did not like these preparations, though spraying the soil enhanced growth and the analysis showed encouraging results. Now we had a problem because prior to sowing we had to stir three different preparations for one hour each.

So we looked for a more time-saving solution: we stirred a mixture of horn manure, horn eggshell and horn basalt for one hour. By applying the three preparations and the mixture, new trials were performed, together with control plots. For several years we used five crop species: spinach, lettuce, oats, beans and radish. The results of these investigations were against the mixture.

Research with hourly ground substances showed a new way. We chose cow-pats as a base material, added chicken eggshell and ground basalt, and dynamized the mixture by turning it over in a circular movement for one hour. Then we put it into a barrel from which the bottom was removed, and which was dug into the ground, and added the five compost preparations, one gram each and ten drops of the valerian preparation. (At the same time the same procedure was performed but the compost preparation was added prior to mixing. This method did not prove to be good.) After four weeks the content of the barrel was mixed by turning over thoroughly with a spade, and after a further four weeks the cow-pat preparation was ready for use in new trials.

Meanwhile, Matthias Thun joined our team, and the necessary trials were performed on three different test areas during two seasons.

In 1972, fourteen years after we reported on the first experiment about the production, application and effect of the Manure Concentrate Preparation, only a short remark concerning the radioactivity was made. We did not want to cause any panic. Nevertheless the manure concentrate has been accepted worldwide and is used together with other biodynamic preparations.

In 1986 the Chernobyl disaster happened. Radioactivity was measured at many farms. A publication in *Lebendige Erde* showed that biodynamic areas were equally contaminated as all the others. However, in some distinct areas the experts measuring the radioactivity had the feeling that there was something wrong with their instruments. They came back with new instruments the next day but these only worked as expected on the area of the neighbour across the path. Meanwhile employees of a government research institute got hold of manure concentrate, produced after the disaster by different farmers and gardeners in the contaminated region. The results were incomprehensible to them — it was not radioactive except for a very small amount of old Caesium (*old* meant that it did not stem from the Chernobyl disaster

Figure 27. Manure Concentrate Preparation in bottomless barrels.

but from fallout of earlier atomic bomb tests). However they did not give any written confirmation of this to avoid the claim that a means against radioactivity exists. I told them something about 'life-promoting radiation' mentioned by Rudolf Steiner in the context of compost preparations.

I was unable to follow Dr Pfeiffer's suggestions for long, as he soon passed over the threshold of death. I owe a lot concerning this topic to another old friend. He arranged all the analyses but did not want to be mentioned, although the results were of great interest to him. The last question I asked him was about the difference between diabase and basalt. He did not answer promptly, but a few days later showed me very thin cuts of diabase and basalt. The texture of diabase looked like rolled cow horn and basalt showed a wonderful crystal structure from the 'self-stage of the Earth' (so called by Rudolf Steiner). This was the legacy of this old friend and our last discussion before his death.

Further effects of Manure Concentrate Preparation

The application of Manure Concentrate Preparation stimulates micro-organisms in the soil, but Mr Dörr (Rheinbischofsheim/Germany) reported:

Figure 28. Harvest with a combine harvester at the test field 'Hassenrodh'.

> For several decades I have worked the land following biodynamic methods and have fertilized the meadows with compost, but the meadows had soil compressions and were always wet so it was difficult to make really dry hay. However, since I applied Manure Concentrate Preparation the meadows are more permeable and let water through. In addition more herbs are growing and the hay dries very well. The hay quality is better and thus also the milk yield.

This indicates a further use of Manure Concentrate to farmers.

Horse manure concentrate

Some farmers have difficulty achieving a good growth of fungi in heaps of pig and bull manure. In such a situation we recommend taking horse manure to make Manure Concentrate Preparation when the Moon is in Virgo. When it is ready the stirred Preparation is sprayed while piling up the compost heap, or one can spray a small amount every day or depending on the way of mucking out, one can use a watering can to pour it on to the manure in the stable once a week. A stirred Preparation can be used for four days and then a new one has to be stirred.

Figure 29. Manure from Fruit days, manure from Aries days is rotting very badly.

Figure 30. Well-shaped cow-pat used to make the Preparations.

The cow-pat

The trials made so far have shown that at the moment of excretion the cow-pat takes in and retains cosmic forces. We know nothing about their effect in the animal organism, but we know that the moment of milking also determines the intake of cosmic forces. Milk leaving the warmth of the animal is open to cosmic forces and the same seems to be true for the cow-pat.

This cow-pat with its cosmic forces, is needed to make the Horn Manure Preparation and the Manure Concentrate Preparation. So we asked which aspect would be the best to produce the preparations and which cosmic forces should be contained in the cow-pat. Many people found that, despite careful production, a Preparation failed and could not be used.

The analysis of the fresh manure is also important. We asked the farmers, who had participated in the manure trials, to collect fresh manure again. Winter was the most suitable time since the same food was needed as well as one person to care for the cows as described earlier. Both farmers were willing to take part. Samples of the animals' morning manure were taken and immediately frozen to ensure comparable conditions.

The analysis was performed in the laboratory of Dr Balzer. Since the food was the same the values should not differ much. However, this was not the case as can be seen in the following tables:

Figure 31. Total nitrogen content of fresh manure (Wedig von Bonin).
Figure 32. Phosphorous content of fresh manure.
Figure 33. Calcium content of fresh manure.
Figure 34. Magnesium content of fresh manure.
Figure 35. Potassium content of fresh manure.

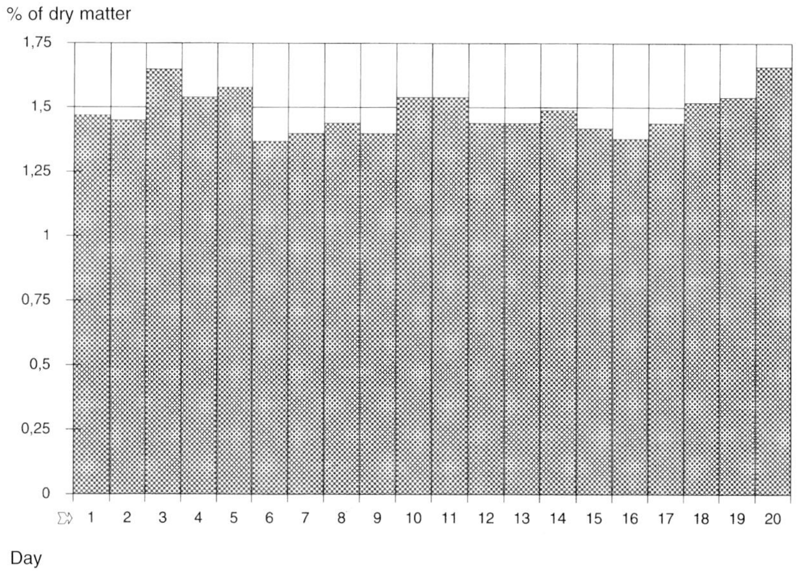

Figure 31. Nitrogen.

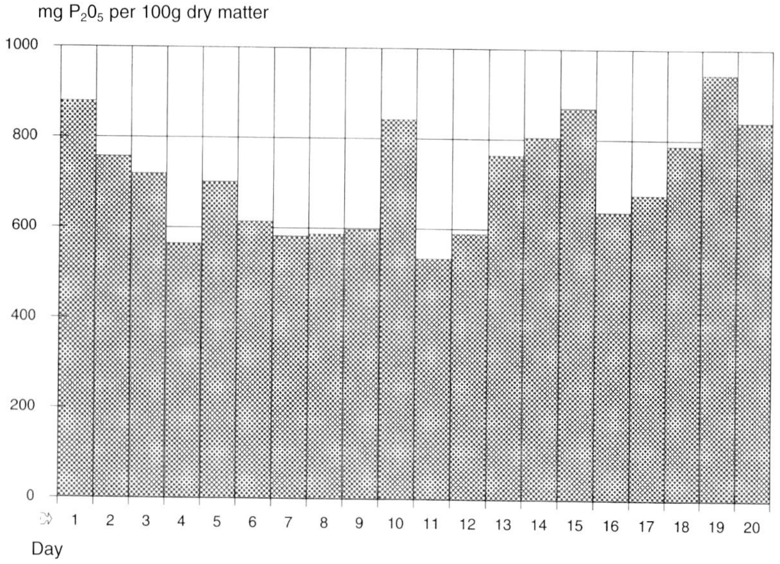

Figure 32. Phosphorus.

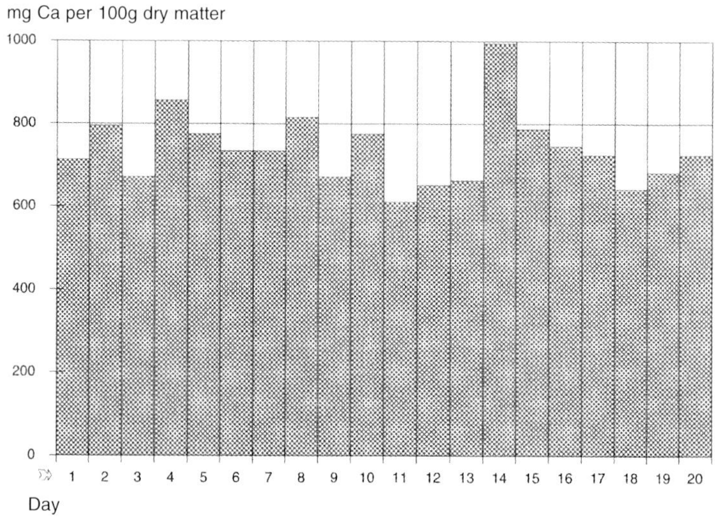

Figure 33. Calcium.

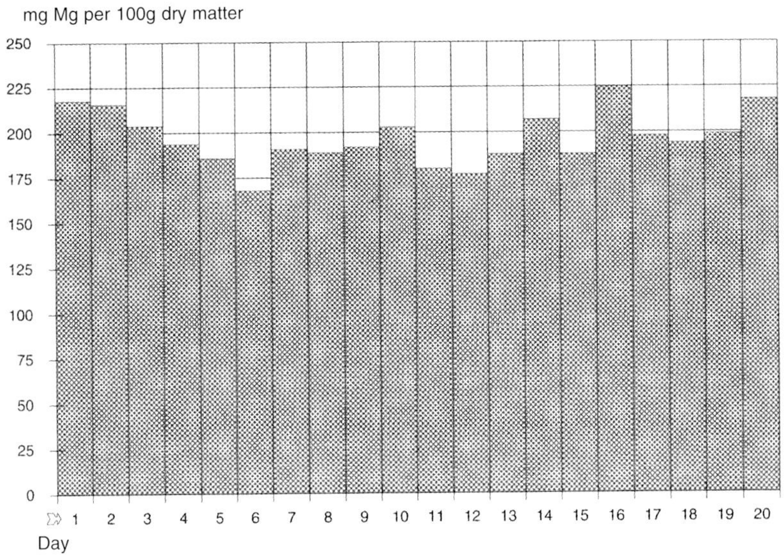

Figure 34. Magnesium.

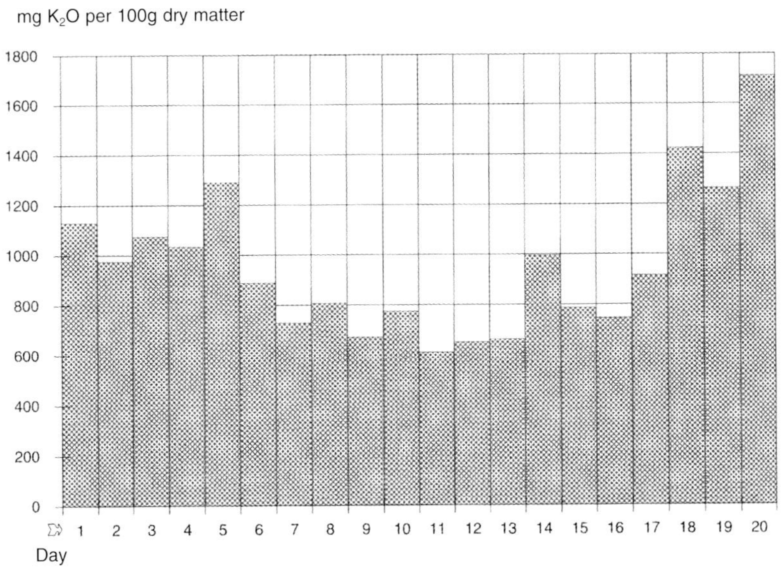

Figure 35. Potassium.

Production of Manure Concentrate Preparation (Wedig von Bonin)

Wedig von Bonin produced Manure Concentrate Preparation during various aspects of the sidereal Moon. The cows were fed with roughage. During the trial the Sun was in Leo and during some repetitions in Virgo. In all cases the preparations were put into bottomless barrels that had been sunk into the ground, treated with the compost preparations and left for four weeks. Then they were turned over with a spade and left until the following spring.

We planned trials in Dexbach and needed areas which had had the same previous crops. For a total of nineteen of Wedig von Bonin's different Manure Concentrate Preparations our test field Unteres Hassenrodh was large enough. The earlier crop was Phacelia. The length of the field was sufficient for twenty-one plots of size 7.5 x 15m. Then nineteen vessels and enough people were required to stir the Preparations for twenty minutes. The plots were sprayed and after each spraying the equipment had to be cleaned. The field was then harrowed and divided into plots again. The same procedure was required for the second and third spraying. It involved a lot of work.

Since Phacelia had been grown on Unteres Hassenrodh leguminosae should be grown the next year. We decided to sow peas, lentils and beans. They were sown lengthways with a sowing machine. Then we stirred Horn Manure Preparation of the same origin in a big wooden barrel. The Manure Concentrate Preparation was sprayed on to the total length of the plot, to a width of 7.5 metres; a width of 2.5m of each plot was left with only Manure Concentrate while the remaining 5 m was sprayed three times with Horn Manure. When the shoots came up the 5m zone was divided into two and the last 2.5 m was sprayed with Horn Silica which had been prepared and sprayed during the same aspects as the Manure Concentrate Preparations.

We had to use a shield to spray Horn Manure and Manure Concentrate Preparations, as our co-workers feared complications due to the strong wind and the small size of the plots (see Figure 38). We had to spray Horn Silica fifteen times. We started stirring the Preparation half an hour before sunrise. One hour later it was ready and we went to the field and sprayed it. We have been performing spraying trials for more than thirty-eight years and have never had a situation preventing the spraying of a stirred preparation. When the

Figure 36. Eighteen barrels containing Manure Concentrate at Wedig von Bonin's farm.

Figure 37. Testing the Manure Concentrate: it is ready for use.

Figure 38. A shield was used for spraying at the test field Unteres Hassenrodh.

Figure 39. A trial with sixty variations of peas and lentils respectively.

Figure 40. Flowering peas.

Figure 41. Neutral after-crops of peas (1992) at Thalacker 1993.

	Manure Concentrate	Horn Manure	Horn Silica
Edge to the west	2.57	2.61	3.08
1 = ♋	2.04	2.34	2.49
2 = ♌	3.38	3.54	4.16
3 = ♍	2.27	2.37	2.58
4 = ♎	2.31	2.73	3.09
5 = ♏	2.48	2.73	2.83
6 = ♐	3.52	4.04	4.42
7 = ♑	3.00	3.02	3.15
8 = ♒	2.73	3.08	3.18
9 = ☿ ♈	2.60	3.01	2.55
10 = 0	2.21	2.26	2.51
11 = 0	2.28	2.49	2.73
12 = △ W	2.73	3.24	3.83
13 = ♓	2.60	2.69	3.14
14 = ♈	3.38	3.46	4.28
15 = ♉	2.99	3.18	3.40
16 = ♊	2.78	2.89	3.09
17 = Pg	1.84	2.36	2.26
18 = ♋ II	2.94	3.02	3.17
19 = ♌ II	3.42	3.58	4.12
Edge to the east	3.34	3.36	3.83

Manure Concentrate trial at Unteres Hassenrodh, 1992. Yield of peas in tons/hectare

fifteenth spraying had been done, the workers realized that all of them had been applied in time. However, it should be noted that when planning such trials we omit days when storms are likely, which is usually possible as there tend to be two days having the same aspect.

In the table above the seed from peas in kilograms, from an area of 10 square metres, corresponding to a yield in tonnes per hectare (t/ha) is listed. It shows the effect of the aspect on the Manure Concentrate and also the Horn Manure and Horn Silica Preparations.

The peas were cultivated in 1993 as neutral after-crops (see Figure 41). The success of the applied method was obvious in a good yield and a good power of reproduction.

So far we have reported the first results of this trial. Later the soil of all sixty plots was analysed, but the analysis of the peas' protein content is still missing. We have also cultivated neutral after-crops of lentils and beans. So we still need some more years to evaluate all the results including those with the Preparations of Graf von Finckenstein (see the table on page 96) which are being performed at the same time.

Trials with manure concentrate

To obtain optimal cow manure for the production of Manure Concentrate Preparation, Graf von Finckenstein changed the fodder for several cows to hay. The aim was to make Manure Concentrate during various sidereal Moon aspects and also at some problematic planetary aspects. The other livestock received seasonal green fodder and their manure was also used to make Manure Concentrate. The morning manure was taken as described before. At the beginning of the trial the Sun was in Leo, and later trials were made with the Sun in Virgo.

We planned our trials and chose the test field Thalacker, which had had wheat as an earlier crop. The aim of the trial was the same as described in the previous chapter but now the different fodder was also taken into account. The test field was divided into twenty-three plots of 7.5 metres wide and 20 metres long. We sprayed two thirds of each plot with Horn Manure and one third with Horn Silica on top.

The preparations were stirred near the plots and some developed a lot of foam. There was more work involved in comparison with the trial at 'Unteres Hassenrodh' because Thalacker-Feider is larger and more species could be sown: sunflower, flax, oats, barley, peas, lettuce, beetroot, carrots and potatoes. Otherwise the procedure was the same as for the previous trial. Big differences could be seen during early growth and were amazing at harvest. Carrots, beetroot and potatoes cultivated on plots sprayed with the foaming Manure Concentrate Preparations rotted in the soil despite a very dry year. Other crops developed very well e.g. the oil-bearing fruits.

After harvest we took samples of the soil to the laboratory of Dr Balzer for analysis. As an example of the results, the barley yield in metric tonnes per hectare (t/ha), is summarized in the following table It will still take some years to fully evaluate this work.

	Manure Concentrate	Horn Manure	Horn Silica
Edge to the west	2.53	2.80	3.09
1 = ♌	3.19	3.52	4.29
2 = ♍ Gr.	2.75	2.91	3.24
3 = ♍ H	2.64	3.13	3.96
4 = ♎ H	3.03	3.19	4.34
5 = ♎ Gr.	2.20	3.25	3.92
6 = ♏ H	3.74	4.62	5.17
7 = ♏ Gr.	3.85	5.06	5.39
8 = ♐ H	3.41	4.03	4.46
9 = ♐ Gr.	4.29	4.12	4.95
10 = ♌ H	4.29	4.62	5.28
11 = ♌ Gr.	3.63	4.18	5.28
12 = ♍ H	2.97	3.41	3.96
13 = ♍ Gr.	3.30	3.63	3.96
14 = ♎ H	3.36	3.52	4.18
15 = ♎ Gr.	3.36	3.85	4.13
16 = ♏ H	2.86	2.97	3.58
17 = ♏ Gr.	3.08	3.41	3.80
18 = ♐ H	3.74	4.18	5.06
19 = ♐☿☋	3.30	3.63	3.80
20 = ♃ H	3.25	3.41	4.02
21 = Light	2.75	2.97	3.08
22 = Water	2.97	3.08	3.19
23 = Mix	3.03	3.50	4.29
Edge	2.53	2.86	3.19

Graf von Finckenstein's Manure Concentrate trial at Thalacker 1992. Barley yield (tons/hectare). The suffix Gr. means the manure is from cows fed with green fodder and suffix H means hay fodder. 21-Light denotes Manure Concentrate with manure from a flower day, which was several years old, 22-Water means manure from a Leaf day and 23-Mix means manure of many days.

Figure 42. Trials at Thalacker with 23 different Manure Concentrate Preparations.

Figure 43. Trials at Thalacker with 72 different Manure Concentrate Preparations.

Compost and Compost Preparations

In *Agriculture* by Rudolf Steiner we find the instructions to make certain preparations. The plants needed are flowers of yarrow and camomile, stinging nettle, oak bark and flowers of dandelion and valerian. They are prepared using animal organs.

Compost heaps can be made with all kind of substances and have to be kept moist, by adding water or liquid manure if necessary. The compost preparations are applied finely and the heaps are covered with a thin layer of peat. The preparations have a radiating effect and the peat cover keeps this effect within the compost heap.

The positioning of the preparations

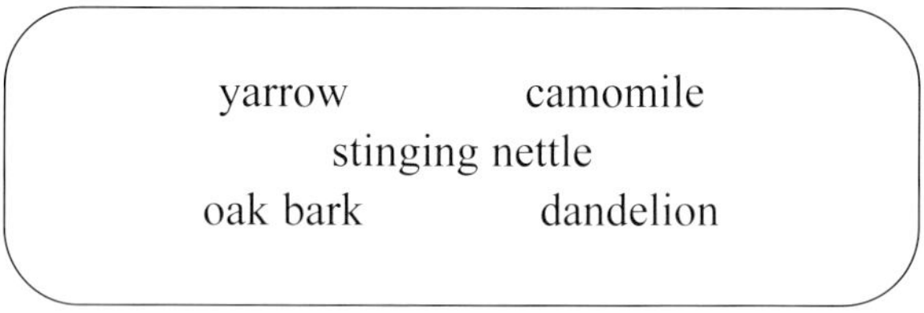

Alongside the heap deep channels are made (0.5 m) with the handle of a rake or a stake at a distance of three metres from the middle (stinging nettle), to the right and left. The preparations are put into them. Afterwards the channels are closed properly. The dandelion preparation is put in the middle on top of the heap. Ten drops of the valerian preparation in ten litres of water is stirred, forming a good vortex for ten minutes in a wooden barrel or an earthenware pot, and then using a watering can poured over the compost heap.

After eight or ten days we can see extended fungal growth indicating the beginning of decomposition.

Even in winter the fungi grow, assuming everything has been done properly, and due to the slight warming caused by the manure itself. The warming up is harmonized by the preparations. For two tonnes of

Figure 44. Volker piling up a compost heap.

Figure 45. After insertion of the Preparations the compost heap is being covered with peat.

Figure 46. Compost with ink caps.

Figure 47. Fungi growing inside the heap.

Figure 48. Mould secreting antibiotics.

Figure 49. Red compost worms appearing after ink caps.

Figure 50. Fungi grow even in winter.

manure one gram of each preparation is required. The ink caps grow with the light. At the second stage, decomposition proceeds with the formation of mushrooms inside the heap. If the temperature there is too high, dry mould develops causing stagnation, probably due to an antibiotic effect. Later on earthworms and red compost worms complete the decomposition. Their excretion is humus-like. The compost heap with the preparations should remain for one year and in autumn be spread on to the vineyard.

Compost trials 1989

Farmers informed us that manure heaps formed from the daily manure decompose differently, so we performed small-scale trials over one month.

We took horse, sheep and poultry manure. Every day the same food was fed to the animals. The morning manure had been collected only if the Moon had the same aspect as it had at the last evening feeding. The manure, a shovel of soil and 20 grams of dry Manure Concentrate Preparation was mixed and subsequently divided into four parts then put into different vessels:

1. Mitscherlich vessels (see Figure 51, third row)
2. Root-boxes (see Figure 52 and 53)
3. Clay pots
4. Cavities in the soil of the greenhouse

All four were kept at the same temperature in the greenhouse. A well-defined amount of water was added three times a week. In the case of the Mitscherlich vessels, liquid seeping out was replaced once a week.

It was noted that with the Moon in Aries a long-haired mould appeared after ten days. The growth of small ink caps started very differently.

After every completion of a lunar phase the seeping liquid was measured and amounts of up to one litre were found. In some cases the whole amount of water was consumed, e.g. when the Moon was in

Figure 51. Compost trial 1989 in the greenhouse.

Figure 52. Root-boxes with manure without fungi formation.

Figure 53. Root-boxes with manure and a good growth of fungi.

Leo. It happened twice, firstly when the Sun was in Scorpio, and secondly when it was in Sagittarius. In both cases ink caps grew which quickly consumed all the water.

In April tomato plants 20 centimetres high were planted on small amounts of the manure, and then developed in a different way. The final height was between 125 and 250 centimetres and the weight of the harvested tomatoes ranged from 800 grams to 2700 grams per plant. The contents of the Mitscherlich vessels had been mixed with 50% garden soil and then aubergines were planted. Soon big differences in growth were obvious.

Chemical analysis also revealed differences. Concerning ground water problems throughout Europe and the conditions imposed on compost plants in these first trials could be pioneering for further experiments. Concerning the question of composting our long-term experience confirmed that a rapid formation of fungi is the most important process.

Now we come back to the question of farmers, who suspected cos-

Figure 54. Newly transplanted aubergines.

Figure 55. Aubergines on manure of different aspects.

mic effects on the daily manure heaps and their decomposition. These trials clearly showed that all living processes on Earth pass in harmony with cosmic rhythms more than was previously believed. For Rudolf Steiner it was self-evident:

> That is why the knowledge of today is not real because most of what happens on Earth would not happen without cosmic forces. As the scientist of today does not speak about these cosmic forces he does not speak about reality. He does not care about reality. Even in the smallest parts seen with a microscope cosmic forces are living in addition to earthly forces. And without taking into account cosmic forces we do not have reality.
>
> (*Mystery Knowledge and Mystery Centres*).

Compost trials 1990

Our compost trials in 1989 confirmed farmers' assumptions that manure is related to the constellations, and the farmers were willing to perform new more relevant trials. This meant that the following conditions were required:

— The same food every day for a certain period
— The same person had to feed and care for the animals
— Only one person milked the cows
— The daily manure was put on areas where a compost heap had never been placed
— Daily observations of the temperature, fungi and seeping water were required
— In November the compost and the layers below were analysed

In Figure 58 the compost heaps of the Grönwoldt family at Tinnum, Isle of Sylt, Germany are shown. Sixty manure heaps were set up during two lunar phases.

In Figure 59 the manure heaps of Wedig von Bonin at Schashagen/Germany are shown. Over thirty days compost heaps were set up and treated. Big differences were visible on the outside of the heaps and everyone was curious to know about the analyses in November.

Figure 60 shows Dr Balzer taking samples at Tinnum. Samples were taken not only for analysis but also for planned planting trials. Subsequently the compost was removed and samples from the soil (at a depth of 25 centimetres and also at a depth of 25 to 50 centimetres) were taken from twenty patches, totalling three samples of each of the sixty heaps. In Dr Balzer's laboratory seventeen different values were measured, but here we are only showing two tables regarding the total nitrogen content of dry matter as a percentage, and the nitrate (NO_3) content per 100 grams dry matter depending on the days in March and April 1990. There were large variations.

Manure samples were mixed with 50% soil and put into clay pots of 14 centimetres in diameter, which were buried in the soil, with the top at the same level as the surface, at the test field 'Heckacker' in November. The following spring the content of the pots were mixed again and then spinach was sown. Figure 61 shows the different growth of the spinach plants.

A small amount of ten tonnes per hectare (10 t/ha) every year is better than 40 t/ha of fresh manure every 3 to 4 years. The mature compost contains 1.9 to 2.0% nitrogen, but fresh manure only contains 0.52%.

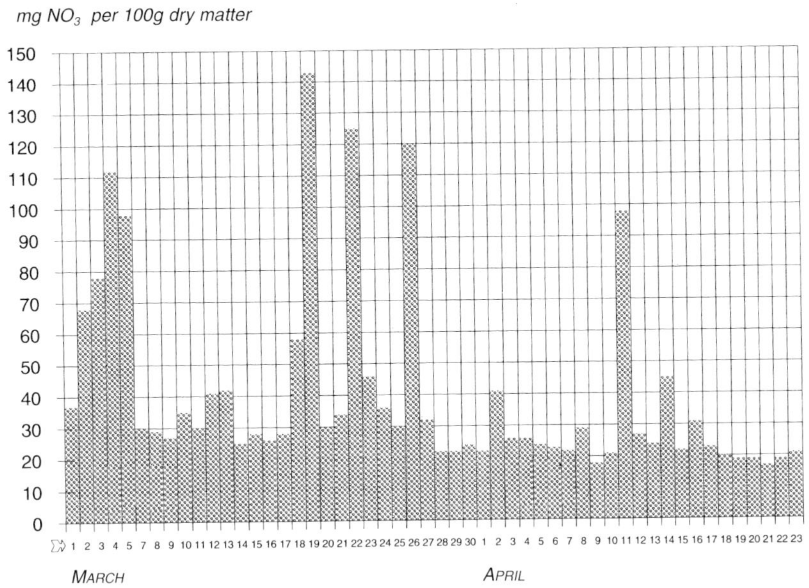

Nitrate content of manure: 1990

Figure 56.

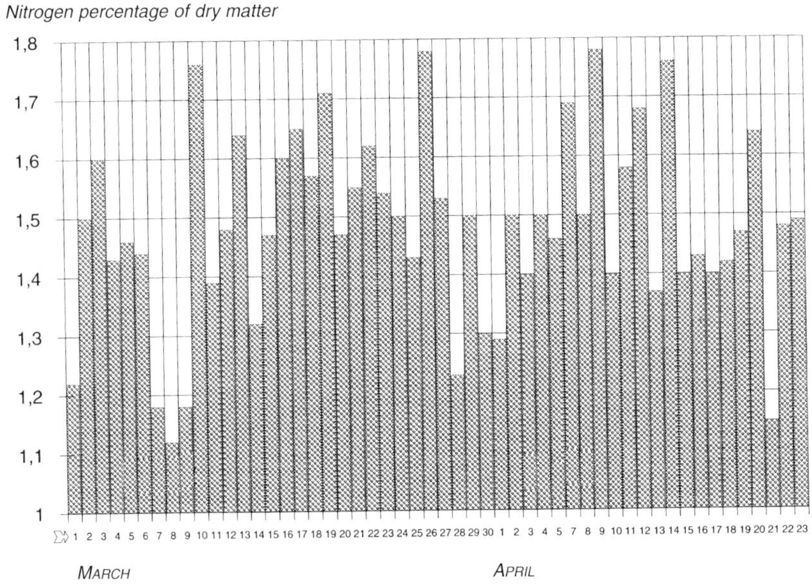

Total nitrogen content of manure

Figure 57.

Figure 58. Manure heaps of 60 days at the Grönwoldt family farm.

Figure 59. Manure heaps of 30 days at Wedig von Bonin's farm.

Figure 60. Dr Balzer taking samples at the Grönwoldt's farm.

Figure 61. Pots with spinach grown on 90 manure samples.

Manuring of gardens

Gardeners often ask how much manure is required. The same amount should be applied to farming and wine growing: about 5 kilograms of dry compost, or a ten-litre-bucket-full is sufficient for an area of 5 square metres. By taking a flat filled wheelbarrow containing 40 kilograms and adding two buckets full we have 50 kilograms — enough for an area of 50 square metres. This amount applied every second year manures the soil very well.

Manuring of cereals

Concerning the quality of seed it is important to consider last year's manuring as well as the time of harvest. If half-rotted manure was applied it resulted in a higher yield of straw and in the following year rye showed a tendency to fungal attacks. The best guarantee for high-quality seed is to use completely humified compost, spray Horn Manure on Fruit days prior to sowing, and then to spray Horn Silica three times also on Fruit days — firstly at the two-leaf stage, and then every nine days on Fruit days — and to harvest on favourable Fruit days.

Figure 62. A wheelbarrow is used to measure the amount of compost.

The Preparations

Horn Manure Preparation

This biodynamic Preparation 500 connects the plant much better with the Earth. When it is sprayed at sowing time the plant forms a better root system and can supply its upper part much better with earthly substances and forces. It works best of all when the soil is moved again after spraying. It should also be used when transplanting and above all it is important to apply it when one cannot use planting time.

Horn Manure Preparation should only be applied when sowing or transplanting. It has primarily the task of enabling plants to orientate and find a connection with the soil. It should be sprayed three times during the preparation of the seedbed. It is advisable to mount the spraying equipment in front of the tractor, and so spray while soil preparation is done. Only one additional hour is needed to stir the preparation. If Horn Manure preparation is of high quality, it can be sprayed up to four hours after it has been stirred. Then a new one has to be stirred.

Horn Silica Preparation

This Silica Preparation 501 works in harmony with cosmic forces. With regard to the quality of plants one gets the best results when it is sprayed on a particular kind of plant on the appropriate day. This means one sprays cereals on Fruit days, root crops on Root days etc. In some cases one can improve the storage quality by using this silica Preparation: e.g. a final spray for cabbage plants on Flower days helps in their storage. One can improve the quality of carrots by applying this spray in the afternoon on Fruit or Flower days shortly before harvesting them. In the morning the spray works more strongly in the upper part of the plant and in the afternoon it works more on the root region.

Horn Silica Preparation attains the best effect if one begins to stir it before sunrise and begins spraying immediately it is ready.

Figure 63. Midnight at North Cape on summer Solstice.

Figure 64. Twilight at Tromso (2am) on summer Solstice.

Twilight is of great importance. As its application in the far north has been problematic, we have investigated for a long time, up to which latitude it could be used. The editors of the calendar flew to the North Cape at the summer solstice, circled over it several times, and landed in Tromso, Norway (latitude 68°N). As we got off the plane it was twilight, which mattered to us. There were mounds of snow in the streets and flowering snowdrops in the gardens as a first sign of spring, comparable to the situation in Dexbach at Easter. We realized how far north the Silica Preparation can be applied. Later we learned that in the northern hemisphere crops are cultivated up to a latitude of 70°N and biodynamic growers exist up to 68° latitude. This means that to apply Horn Silica in these northern regions one has to start stirring soon after midnight and spray right away so that the forces of twilight, together with those of the preparation can help the plant.

Horn Silica trials: wheat 1988

To study the effect of Horn Silica on the growth of wheat and its baking properties we sowed winter wheat in the autumn of 1988. The year before we had grown potatoes in the same area and had applied 20 tonnes per hectare of mature compost to the potatoes. Therefore, no manure was needed for the wheat. At the same time we worked the soil prior to sowing and sprayed Horn Manure three times. The wheat was sown on September 20. This early date was chosen since we planned three Horn Silica sprayings on each of the four aspects — root, flower, leaf and fruit. Later the field was divided into plots to allow replications and control plots.

When the second leaf formed we started to spray. After sunrise on 15 October we sprayed the Leaf day plots for the fist time. All twelve plots were sprayed twice with the Sun in Virgo, and a third time with the Sun in Libra. The plants developed and showed clear differences, firstly in the colour of the leaves and the following spring in the development of tillers and the formation of ears.

As we had to use the combine harvester, all varieties were harvested on 23 August. After cleaning the wheat was stored in hanging canvas bags and evaluated in 1990.

Since the application of Horn Silica influences the ability of the wheat to swell as well as the gluten content, it is of great interest to the baker how much bread he can make. The following table shows

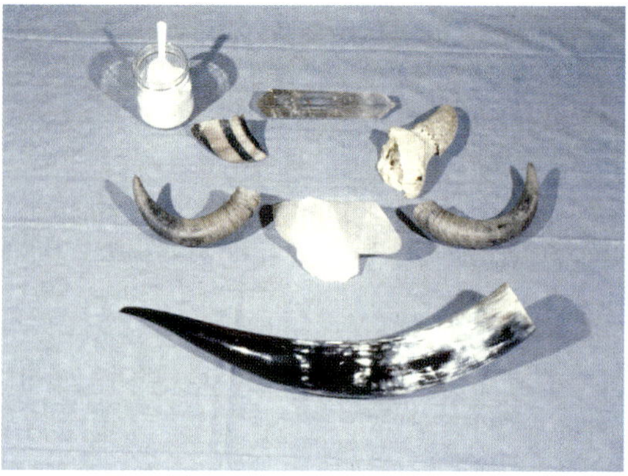

Figure 65. Production of Horn Silica.

the weight of one thousand grains in grams (equivalent to one grain in mg) and the amount of bread that can be baked in tonnes per hectare, evaluated from the grain yield per hectare and the bread yield of our baking trials. We also tested the ability of the wheat to germinate and found that the application of Horn Silica at correct aspects increased not only the grain yield and the good bread-making properties but also the regenerative forces.

	Brown with awns		'Okabi'		Summer wheat	
Application of Horn Silica	mg	Bread (t/ha)	mg	Bread (t/ha)	mg	Bread (t/ha)
Control plot	54	75.6	40	56.4	32	51.4
Leaf days (AK)	58	84.2	44	61.6	34	61.2
Flower days (LK)	62	86.8	45	63.8	35	59.4
Root days (EK)	55	79	42	58.8	34	53.2
Fruit days (WK)	67	93.3	48	67.2	38	67.5

In *Agriculture* Rudolf Steiner states that lime substances in the soil mediate the forces for the regenerative power of the plants which emerge from Mercury, Venus and the Moon. The ability of the plants to nourish human beings and animals is a gift of the outer planets mediated by the silica substances of the soil. It seems likely that by taking cosmic rhythms into account for the application of preparations we can initiate ambient forces to work creatively on the plants.

Horn and Hoof Manure Preparations

Production at various aspects

Since more traditional farmers worked the land biodynamically it became difficult to provide enough horns to make the Preparations (in 1990 and later). In *Agriculture* Rudolf Steiner mentioned horns and hooves. Farmers like the horns polished as gifts and souvenirs. In contrast hooves have been in contact with manure and are not inspiring at all, but the view might change if we take into account that a cow has eight hooves but only two horns. For a trial we needed all parts of one cow and in addition the manure, kept prior to slaughter. Trials were performed on fields and in pots. Figure 66 shows peas (from left to right) from a trial with horn, water, hoof and toe. Unfortunately we have no photographs of the roots. For horn and hoof the formation of root nodules was very good but less so for the Toe Manure. For this reason we used only horns and hooves in further trials. As mentioned we tested Horn and Hoof Manure for more than ten years but not Horn Silica Preparation.

In the course of numerous trials we used also Horse Manure Preparation. In his course on agriculture Rudolf Steiner was asked if horse manure could be put into cow horns. His answer was yes but it would be necessary to wrap the horn with one hair of the horse's mane to bring the forces into play, since a horse has no horns (see *Agriculture* page 85). So we included Horse Manure Preparations in the present trial. Horn and Hoof Manure Preparations were produced at different Moon aspects.

Spinach trials 1991

Despite the limited test area the aim was to have a large variety of crops, root plants, leaf plants, flower plants and fruit plants to see the effect of soil transformation. We chose tomatoes, potatoes, marigolds, borage, spinach, radishes, barley, peas and spinach beet.

We grew spinach in pots to complete the field trials. Figure 71 shows (from left to right) spinach grown on soils sprayed with Horn Manure, nothing, Horse Manure, Hoof Manure. Figure 72 shows (from left to right) spinach grown on soil sprayed with Horn Manure produced at different Moon aspects: Libra, Virgo, Leo, Pisces. No Horn Silica was applied. The results of the field trial with spinach are shown in the table below. We used our own seed; gained from cultivation with the Moon in Aries and the Moon in Pisces. The spinach yield corresponds to the amount of every second plant harvested from five rows per 1.5 metres each 5 metres long. The other plants were left to go to seed. In the following season we grew after-crops from these seeds. Prior to sowing we sprayed Horn Manure three times. Later we sprayed Horn Silica three times. We harvested every second plant again from a row five metres long. The rest was left to go to seed ('Horn cow' and 'Hoof cow' means the preparation was made from dung of several days, the aspects are unknown).

Spinach yields in grams/ five rows 2.5 metres long

Neutral after-crops of spinach (1992)

Production of Horn Manure	seed (Aries) leaves	seed (Aries) seed	seed (Pisces) leaves	seed (Pisces) seed
Leo	3500	960	3900	620
Virgo	3800	785	4300	510
Libra	3300	650	3600	400
Pisces	4800	880	5200	520
Control Plot	3100	520	3400	380
Horse	3800	920	4100	610
Horn-cow	3580	690	3650	430
Hoof-cow	3510	650	3850	420

Production of Horn Manure	seed (Aries) leaves	seed (Aries) seed	seed (Pisces) leaves	seed (Pisces) seed
Leo	5300	756	6400	710
Virgo	4150	630	5500	615
Libra	6000	465	4500	450
Pisces	3300	360	3200	345
Control Plot	4850	630	5500	650
Horse	3600	465	3800	426
Horn-cow	3900	510	3900	450
Hoof-cow	4000	600	4200	450

Figure 66. Peas, first trial with Horn and Hoof Manure: Horn, Water, Hoof, Toe (from the left).

Figure 67. The dung from these cattle was used for most trials.

Figure 68. Horse dung was used to make Horn Manure Preparation.

Figure 69. Horn and Hoof Manure Trial, spinach (1991).
Figure 70. Horn and Hoof Manure Trial, peas (1991).

Figure 71. Horn and Hoof Manure trial, spinach in pots: 500 Horn, a mix of several horns, Water, 500 Horse, 500 Hoof (left to right).

Figure 72. Horn and Hoof Manure trial, spinach in pots: 500 Libra, 500 Virgo, 500 Leo and 500 Pisces (left to right).

As already mentioned no Horn Silica was applied in this trial to prove the effect of Horn Manure Preparation. We found that some plants behaved very differently. Radishes (*Raphanus sativus radicula*) produced large crops and a large amount of seeds. In the case of Horse Manure the crops had a size comparable to that of white radishes. We observed this in earlier trials with Manure Concentrate Preparations of horse manure. So it confirmed the fact that special preparations can have very selective effects. The same Preparation also had a big effect on oats and had also been found in earlier manure trials with horse manure.

Wheat trial with Horn Manure 1992

The wheat trial was sprayed three times with the same preparations as for the spinach trials in 1991. No other manure was applied prior to sowing summer wheat. Horn Silica was sprayed once except on the control plot. The following yields were harvested:

Horn Manure Leo	5.76 t/ha	Horse Manure	5.20 t/ha
Horn Manure Virgo	4.73 t/ha	Horn Manure cow	4.92 t/ha
Horn Manure Libra	4.72 t/ha	Hoof Manure cow	4.90 t/ha
Horn Manure Pisces	4.93 t/ha	Control & Horn Silica	4.52 t/ha
		Control plot	4.24 t/ha

Figure 73. Spinach after-crops 1992 at Heckacker.

Figure 74. Horn Manure trial: wheat 1992.

'Horn Manure cow' and 'Hoof Manure cow' was made with the dung of several days but the aspects remain undefined. Spraying Horn Silica to the control plot increased the yield by 0.3 t/ha. Assuming a corresponding increase due to Horn Silica for all other plots the following values of yield increase can be evaluated:

Horn Manure Virgo	0,21 t/ha
Horn Manure	0.39 t/ha
Hoof Manure	0.39 t/ha
Horse Manure	0.68 t/ha
Horn Manure Leo	1.26 t/ha

This represents a considerable yield increase for homeopathic treatment. Actually no additional work has to be done to make the preparations at certain aspects.

Spinach trial 1992: Horn and Hoof Manures

In 1992 we continued the trials with various manure preparations that were started 1991 (see Horn and Hoof Manure Preparation). Our test area was limited for the horn and hoof trials, because other trials had been made in the larger fields. Before new trials can be made in the same field a uniform compensatory planting must be made for a year. Therefore we could not make tests during all constellations of the zodiac. However, we decided to make new Preparations during constellations allowing some repetitions. All Manure Preparations were sprayed three times prior to sowing, and we added spraying Horn Silica three times at intervals corresponding to the Preparations. This means if we have a Leo Manure Preparation we sprayed Horn Silica on Warmth-Fruit days etc. To compare with last years results, the spinach yield of every second plant harvested from an area 5 times 1.5 metres with 5 rows are presented opposite:

Figure 75. Spinach after-crops 1993 at Thalacker.

Horn Manure	plus 3 times	Horn Silica
Horn Leo	4.0 kg	5.2 kg
Hoof Leo	3.6 kg	5.1 kg
Horn Virgo	3.5 kg	4.4 kg
Hoof Virgo	3.3 kg	4.2 kg
Horn Libra	3.4 kg	3.9 kg
Control plot	2.8 kg	3.6 kg
Horn Scorpio	4.5 kg	5.9 kg
Hoof Scorpio	4.4 kg	5.6 kg

The results clearly show the cosmic effect at the moment of excretion and the production of Preparation, but also reveal the effect of the lunar aspect during the application of Horn Silica. The analysis revealed the lowest nitrate content and the highest sugar content for the two Scorpio constellations. The analysis of the after-crops grown in 1993 yielded the same results.

Spinach trial with Horn Silica and hoeing

On April 22 spinach was sown. Then the field was divided into plots; three plots for each variety. Hoeing and spraying was always done in the morning.

A = hoeing on Leaf days	AK = hoeing and Horn Silica on Leaf days
W = hoeing on Fruit days	WK = hoeing and Horn Silica on Fruit days
E = hoeing on Root days	EK = hoeing and Horn Silica on Root days
L = hoeing on Flower days	LK = hoeing and Horn Silica on Flower days

Horn Silica was sprayed three times in the morning. At that time we did not know that it could only be applied successfully on Flower days very early in the morning. In this trial its application on Flower days set off a depression in the plants leading to hardening, loss of taste and hinders the substance formation. Consequently the analysis revealed a higher nitrate content. Today we know at which aspect and time the Horn Silica Preparation has a promoting effect on the plant. When we sprayed Horn Silica on oil-plants during daybreak it had an especially positive effect. The spinach yields harvested are depicted in the following graph.

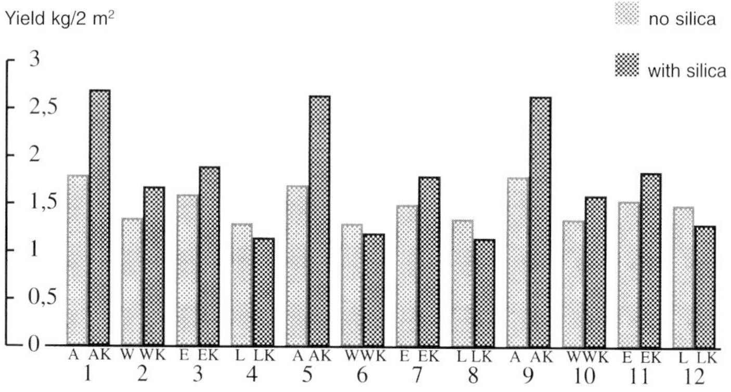

The mean spinach yields in kg/m²

A	1.83	AK	2.72
W	1.23	WK	1.72
E	1.54	EK	1.75
L	1.35	LK	1.18

In all cases Horn Silica led to an increase in yield and quality except when applied on Flower days. Similar observations were made for other plants.

Figure 76. Spinach trial at Gisselberg 1971.

Lettuce and lambs lettuce trials 1992

In 1992 Horn and Hoof Manure trials were continued with trials of lettuce and lambs lettuce. For spinach the best results were obtained with Horn Manure Preparation produced with manure excreted when the Moon was in Scorpio and Virgo. However a closer look revealed that on the day the Leo Horn Manure was produced a special influence of Uranus predominated. A strong influence of Neptune was present on the day the Virgo Horn Manure was produced. On the days the Scorpio Horn Manure was made, no planetary aspects played a role. This means we have to take planetary aspects into account when preparing Horn Manure, as we do for sowings. Therefore, new trials are necessary until we achieve final results. The yield of lambs lettuce and lettuce in kilograms per ten square metres (equal to metric tonnes per hectare, t/ha) is depicted in two graphs. For both crops the additional application of Horn Silica (501) increases the yield in all cases. The highest yields were obtained when Virgo Horn and Hoof Manure was sprayed.

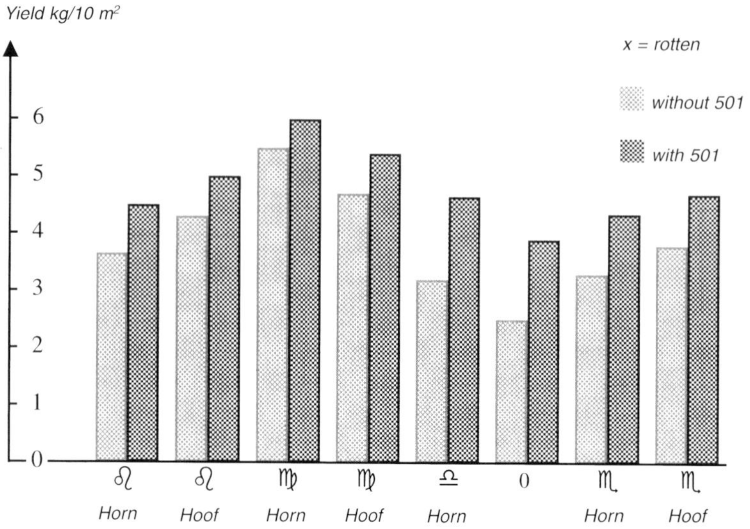

Figure 79. Lambs lettuce yield 1992.

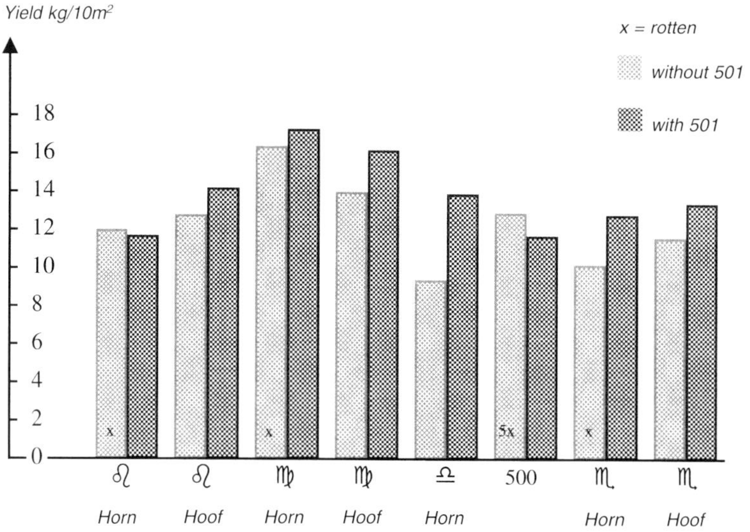

Figure 80. Salad yields 1992.

Figure 77. Lambs lettuce (Horn Manure trial 1992).

Figure 78. Lettuce (Horn Manure trial 1992).

Further Horn Manure Preparation trials 1992

In 1992, to investigate the effect of various Horn Manure Preparations (see also the preceding chapters) we sowed carrots, beetroot, lettuce, lambs lettuce, spinach, radish, cucumber, beans, phacelia, potatoes and onions.

With regard firstly to potatoes and onions: for sowing and spraying Horn Silica both crops were regarded as root plants although they are not real root plants. Both plants were less sensitive to the manure treatment than leaf and fruit plants.

The following potato (Grata) yields in kilograms (kg) of 50 plants were harvested:

Horn Manure		plus 3x Horn Silica
Horn Leo	35.0 kg	37.0 kg
Hoof Leo	27.0 kg	35.0 kg
Horn Virgo	32.0 kg	37.5 kg
Hoof Virgo	25.5 kg	34.5 kg
Horn Libra	31.0 kg	30.0 kg
Control plot	23.5 kg	28.5 kg
Horn Scorpio	29.0 kg	34.0 kg
Hoof Scorpio	29.5 kg	33.5 kg

Figure 81. Onions in August harvested the previous season. Onions harvested on Leaf days (AK), Flower days (LK), Root days (EK), Root day plus Warmth trine (EK DW), K denotes the application of Horn Silica (from right to left).

Figure 82. Clamp for goat food.

Figure 83. Cucumber after-crops of the Manure trial 1992.

Figure 84. Manure trial with runner beans.

Figure 85. After-crops of French beans and cucumbers of the Manure trial 1992.

Onions ('Stuttgarter') were cultivated by applying the same Preparations as for potatoes. They were grown in rows with a plant distance of 10 centimetres and a row distance of 30 centimetres.

The following yield (kg) from a five-metre-long row was harvested:

Horn Manure		plus 3x Horn Silica
Horn Leo	7.3 kg	6.6 kg
Hoof Leo	6.6 kg	6.8 kg
Horn Virgo	6.4 kg	7.6 kg
Hoof Virgo	6.8 kg	7.5 kg
Horn Libra	5.9 kg	5.9 kg
Control plot	4.2 kg	4.9 kg
Horn Scorpio	4.8 kg	5.4 kg
Hoof Scorpio	5.3 kg	5.7 kg

Obviously potatoes and onions do not favour the Libra Horn Manure Preparation and the Horn Silica sprayed on Flower days brought no yield increase. Only onions harvested on Root days kept well. If harvested on Flower or Fruit days they began to sprout and those harvested on Leaf days began to rot early (see Figure 81).

French beans (yellow) were treated in the same way as potatoes and onions and the following yield (kg) from a five-metre-long row was harvested:

Horn Manure		plus 3x Horn Silica
Horn Leo	5.0 kg	9.6 kg
Hoof Leo	4.8 kg	7.1 kg
Horn Virgo	6.0 kg	4.5 kg
Hoof Virgo	3.8 kg	3.7 kg
Horn Libra	4.3 kg	6.4 kg
Control plot	4.3 kg	4.8 kg
Horn Scorpio	4.5 kg	5.7 kg
Hoof Scorpio	4.5 kg	5.1 kg

Beans are Fruit plants. The same amount was harvested for both Leo Manure Preparations. The application of Horn Silica on Fruit days meant that beans could be harvested until the autumn. Three plants of each variety remained for seed to give neutral after-crops next season.

Cucumbers ('Delikatess') were also treated in the same way as the crops described already. We harvested the following yield (kg) from a five-metre-long row and from plants 10 centimetres apart:

Horn Manure		plus 3x Horn Silica
Horn Leo	45.8 kg	52.3 kg
Hoof Leo	51.0 kg	55.4 kg
Horn Virgo	40.1 kg	44.7 kg
Hoof Virgo	47.8 kg	49.6 kg
Horn Libra	48.8 kg	57.3 kg
Control plot	38.6 kg	42.4 kg
Horn Scorpio	39.1 kg	38.4 kg
Hoof Scorpio	34.2 kg	38.4 kg

Cucumbers are also Fruit plants and like the preparations of Leo and Libra made with horns and hooves. Horn Silica spray has a stimulating effect but don't spray it on Leaf days. Cucumber seed was grown to cultivate neutral after-crops and to study the influence of the various preparations on the forces of reproduction.

The use of stinging nettle

The stinging nettle, after it has undergone a special treatment, is used as one of the biodynamic compost preparations. Apart from this various other ways of using it have been developed, either to assist in the biological control of pests and diseases or in the enhancement of plant growth.

The 24-hour extract

For this fresh stinging nettle is used, but without the roots. It does not matter if it is already in flower, but it should not have gone to seed. One kilogram (2.2 lbs) of the plant is put into a wooden, clay or enamel container and 10 litres ($2^1/_4$ gallons) of cold or luke-warm water is poured on it. This is left standing for 24 hours. When sieved the juice is used as a spray in cases of attacks by any kind of larvae or caterpillars. The treatment should be repeated twice within the space of a few hours.

Stinging nettle manure as a growth stimulant

The same preparations are made as for the 24-hour extract using the same proportions, but this time the liquid is left standing until the leaves have at least decomposed. This may happen in 3–4 days, but it may also

Figure 86. Matthias spraying Horn Silica at Rauischholzhausen.

Figure 87. Gertrud stirring preparations at the test field.

take some weeks and depends on the outside temperature. We use the term 'manure' because the smell of this liquid is similar to that of animal manure. It has a potent effect on growth and must be diluted in the proportion 1:10. If, for example, growth has been inhibited by unusually cold weather, which sometimes causes subsequent attacks by aphids, then this dilution can suitably be used. Spraying the soil either towards evening or in the early morning and watering well a few hours later can help plants, which have suffered. Observation of roses, soft fruit, fruit trees and such like has shown that this treatment, also repeated twice within a short period of time, will, through the fact that the saps begin to flow again more vigorously, cause aphids and often also fungal attacks to disappear.

A general tonic for plant growth

The principle of this manure is the same as described for the 24-hour extract, and its preparation is that as described for the growth stimulant, only now the actual manure constituent is less. An amount of $^1/_4$ litres ($^1/_2$ pint) of nettle manure mixed with 10 litres ($2^1/_4$ imperial gallons) of water is used to water crops like tomato, cucumber, spinach, and cabbage. A spray can be made by using $^1/_2$ litre (just under a pint) to 10 litres ($2^1/_4$ imperial gallons) of water and stirring it for 15 minutes before spraying it through fine jets on to the growing plants. Potatoes respond well to this treatment; also soft fruit showed in the following year the result of having been sprayed after harvesting. There should, however, never be more than three successive applications as otherwise the quality of the produce might begin to suffer. This shows in a lessened ability to keep when stored, and also in reduced germination of the seed; both these results are particularly noticeable if the 'manure' is used in a more concentrated form than has been indicated here.

Stinging nettle compost

If compost is made from stinging nettle only, the most excellent soil results, which is particularly suitable for growing delicate crops and for treating roses and strawberries. Only a really successful pine needle compost comes anywhere near it in quality.

PART THREE

Planting, Sowing and Harvesting

The Grouping of Plants

Plants whose edible part is the root are influenced by Earth element. Examples are: carrots, fodder beet, radishes, celeriac and potatoes. These and similar plants do best when they are sown and tended on Root days. When the sowing and care of flowering plants is carried out on Flower days we see an enhancement of tillering and branching as well as a profuse display of flowers.

Leaf plants are plants that have their main productive element in the leaf region as is the case with lettuce, spinach, leeks, lambs lettuce and the cabbage family but kohlrabi and curly kale also react well if sown on Leaf days.

Fruit plants are regarded as those plants that produce edible fruits such as tomatoes, cucumbers, beans, peas, grains, oil-bearing fruits, certain herbs as well as fruit, berries and grapes. These are best sown and cared for on Fruit days. These days also produce particularly good quality seeds.

Every time we hoe or cultivate the soil, however superficially, we encourage new cosmic forces to be active in the ground. If we use the favourable days for a particular type of plant we help in their development. If we use unfavourable days we hinder it. Such growth hindrances leave the plant prone to attack by pests.

During what in the Calendar is described as 'Planting Time' one may also sow, always of course according to the main organ of the plant. It is also recommended to transplant plants at planting times so that they quickly form new roots in their new position. When doing this one should again take into consideration the main type of plant, e.g. lettuce on Leaf days.

Crop Rotation

The ideal rotation picture lies in the plant itself. It lives in a certain fivefoldness and in the course of a few years it wants to experience itself in a succession of possibilities: root, flower, leaf, seed and fruit. We find this series in the cosmic orientation of impulses mediated by the zodiac. We can experience rotation as an organism and the single part of it as an organ. If we compare this with a human being we can perhaps speak of a system of organs. In *Agriculture* by Rudolf Steiner we can find a reference that allows us to see a model form of rotation. This link with the human being can be seen as follows: the human being is linked to the plant in so far as matter works from different areas — the root is reflected by the brain system; the leaf as reflected in the human lung system; the blossom as reflected in the kidney system; the fruit is reflected in the blood; and the seed in the human heart system. If one looks at the different cosmic rhythms this century, one can find a certain orientation which is astonishing in the conjunctions of Venus with the Sun. We have five such conjunctions about 72° apart and moving only very slowly — only 30° over one hundred years. At present they lie in the following constellations: Taurus-Root-Brain, Sagittarius-Fruit-Blood, Leo-Seed-Heart, Pisces-Leaf-Lungs, Libra-Flower-Kidneys and then back again to Taurus. The sixth conjunction almost coincides with the first, being only about two degrees away. If we place the conjunctions in a diagram with the zodiac we get Figure 88.

It seems to me that if we successfully bore these relations in mind over time, the plant would be able to provide the right nourishment for the appropriate organs of the human body. During a single year one would have the plant divided up into its five basic types in the different areas in field or garden, and in the proper rotation on the individual plot. When Rudolf Steiner speaks to farmers and gardeners about an individual organism, which can also work in time, it seems to me that we can achieve something of this with the right kind of crop rotation.

During the last few decades the trend has been to go over to cereal-dominated rotations because there are fewer and often no animals on

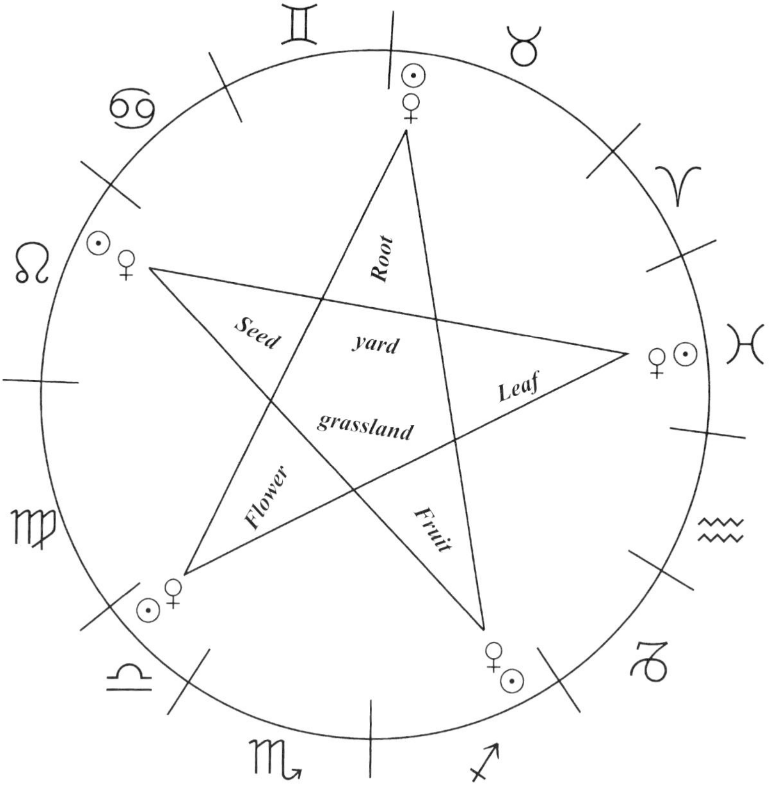

Figure 88.

the farm. As a result, various diseases have appeared affecting leaf or root that call for chemical means of control.

The rotation of farm crops depends also upon soil conditions, climatic situation, animal population and the extent to which machinery can be used, and has, therefore, to be worked out for every farm on an individual basis. For the gardener and smallholder it has often been a problem to achieve good rotation. Some of the plants have a growing season of only a few weeks; so one plant chases the other on the same plot. If the gardener has not written down his rotation beforehand, he is likely to lose sight of his overall scheme and over the course of a year will get into rather a muddle.

The soil becomes exhausted in a one-sided kind of way when plants of the same kind follow each other too soon in the same place. It causes deficiency symptoms to appear in the plant, which weaken it. Nature will step in with other organisms, which come along and do away with weaklings. We may complain then of insect and fungal attacks which we ourselves have caused through our inconsistencies. These pests are often very troublesome to get rid of by biological means. And even if we are successful in restoring the plants to some sort of healthy balance, the quality is, nevertheless, often impaired.

If we look, for instance, at the vegetable members of the Cruciferous family we realize that they appear in the most varied forms. All cabbages hold back the stem forces in their leaves; leaf by leaf is folded over and remains on this level until harvest time. With others the neck of the root expands to produce a fleshy swelling, such as with radishes and swedes. Sprouts set their 'fruit' all along the stem in the leaf axils. Kohlrabi enlarges its stem to a truly delicious vegetable; the flavour is even intensified when the Cruciferous plant is transformed into a cauliflower. Different forces are necessary for each of these various developments, which, on the one hand, make definite demands on the soil, but, on the other hand, also leave residues in the soil that can become troublesome.

At the end of the season it is not only the natural tendencies of the Brassica which are exhausted, but also the forces of the particular 'fruiting organ' as described here. The following season, apart from choosing a different plant family, a different plant 'organ' has also to be developed on that particular piece of ground.

The following may serve as an example of crop rotation for the vegetable garden that has proved successful for a great many years. Failing other neutral crops, old strawberry beds can be taken as a good starting point. After cropping, the bed should be broken up and sown with rye mixed with Persian clover (annual, *Trifolium resupinatum*). Rye is beneficial, as the soil needs grass from time to time.

In the autumn one plants the strawberries shallowly when the Moon is descending (in other words during planting time) and sprays the manure concentrate three times if possible. Naturally different types of plant need their own special compost provided before sowing or planting out. Those areas not needed in the autumn are left fallow for the winter. After 25 years of experiments conducted by Professor von Bogouslawski and Dr Debruck at the University of Giessen, the conclusion was that the traditional way of leaving an area fallow in the autumn brought the best results as regards soil fertility and yields, when compared with more modern methods.

The plan for the first year is as follows:

Plot 1	Plot 2	Plot 3	Plot 4	Plot 5
Green Cabbage	Carrots	Marrowfat Peas	Potatoes	Strawberries
Red Cabbage	Parsnips	Broad Beans	Potatoes	Strawberries
Savoy Cabbage	Scorzonera	Sugar Peas	Potatoes	Strawberries
Cauliflower	Beetroot	French Beans	Potatoes	Strawberries
Curly Kale	Onions	Sweet Corn	Potatoes	Strawberries
Sprouts	Celeriac	Runner Beans	Jerusalem artichoke	Strawberries
Kohlrabi	Leeks	Paprika	or Flowers	Strawberries
Turnips	Spinach Beet	Cucumbers	Jerusalem artichoke	Strawberries
Radishes	Fennel	Tomatoes	or Flowers	Strawberries
Horseradish	Parsley	Cucumbers	Flowers	Strawberries
1st year	2nd year	3rd year	4th year	5th year

In the following year crops can be changed. Spinach, lettuce and endive can be planted as early or late crops. They are not dependent on Fruit days, but should not be planted after each other in the same year. With after-crops always add a little compost.

Sowing and Cultivation Times

By *sowing time* we mean the moment when we place the seed in the Earth. The influence of cosmic forces is intensified when we cultivate the soil at the same time; that means when we prepare a proper seed bed for each individual plant species so that the seed can not only germinate but send its roots down into the Earth easily. At planting time we place the seed in the Earth and in this sense the potato tuber is also a seed.

It should be emphasized when it is a question of sowing and hoeing that cosmic forces enter the soil whenever it is moved or worked, and that they influence, for good or ill, the results of what is done at the time of hoeing or cultivation. If one is prevented by bad weather conditions from making use of suitable times, matters can often be improved by tending the plants later under more favourable conditions.

With regard to hoeing, it is worth remembering that about every nine days the Moon is again in the same triad of forces as it was on sowing day. So if one keeps to these rhythms the cosmic forces continue to be strengthened.

The Earth breathes out in the morning and in again during the afternoon. During wet weather hoeing on the morning of Flower or Fruit days can enhance the evaporation of the moisture in the soil. During a dry period night moisture and dew formation can be increased by hoeing in the evenings of Leaf and Root days, and by spraying with the Manure Concentrate Preparation.

When we sprayed the Horn Manure or the Manure Concentrate Preparation on the paths and the surrounding areas of lawn, towards evening during periods of drought, we found that early the following morning there were delicate veils of mist over planted areas and it was clear that there had been stronger dew formation.

Cultivation of Cereals

For 44 years we have carried out trials with all kinds of cereals to answer various questions. At first we tried to find out which cosmic aspect was the best for sowing cereals in order to get healthy plants and a high grain yield. Later the question of regenerative forces led to trials of growing neutral after-crops. Here we found unfavourable aspects diminishing the ability to germinate. Also several farmers asked if there was a best aspect for spraying the Preparations. Therefore in 1963 we also started to perform trials of spraying at different aspects for all sowings.

Farmers who were growing carrots on a large scale realized that carrots harvested on certain days rotted very quickly. So we performed harvesting and processing trials, and also we carried out compost trials in between. For more than six years we kept goats and carried out trials with milk, butter and cheese. For twenty-one years we also performed trials with oil-bearing fruits. Numerous trials with biodynamic preparations followed.

In addition to the inevitable research work we continued cereal trials. We got twenty-three samples of rye grown at an altitude of 1000 metres and more above sea level to investigate the influence of our methods. It turned out that five species were suitable for propagation and cultivation by Demeter farmers. Similar questions concerning forty species of wheat are still open since the trials are not yet complete.

Finally we asked: which aspects of the growing plants (cereals) reveal the effect of rhythms and preparations.

Favourable and unfavourable aspects became evident in harvest trials. Our wheat never had blight, but growing neutral after-crops of five species harvested at eighteen different times showed blight on five varieties. The same experiment performed with two rye species showed ergot for equal harvest time varieties. So we recommend choosing favourable harvest times for seed production in order to probably avoid work-intensive bathing of seed.

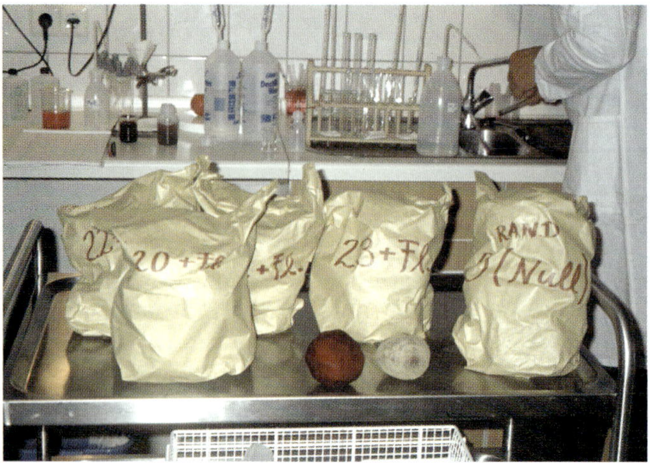

Figure 89. Beetroots prepared for analysis.

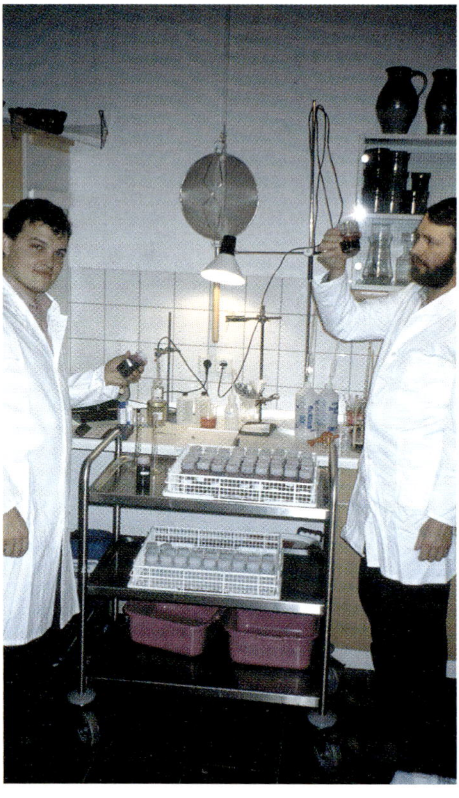

Figure 90. Carrots and cucumbers to be analysed.

Figure 91. Nicolai and Georg analysing plant sap.

Caring for Meadows and Pastures

The course of the year for meadows begins in the autumn. In the past compost made from chaff, wood-ash, weed seed, earth from ditch cleaning, poultry and pigeon manure was spread on the meadows in autumn. Prior to this the rank patches were mowed. The leaves of trees and hedges were also added to this compost. During the winter many molehills appeared, as the winter forces enliven the soil of the molehills. In *Agriculture* Rudolf Steiner mentioned that if something emerges above normal ground level (e.g. compost heaps) it has a tendency to absorb ethereal forces and life energy. This is also true for the molehills. At the beginning of spring farmers flatten and spread the molehills over the meadow by putting down a bundle of twigs, pulling the harrow weighted with some heavy stones upside down on top. That was the only fertilizing. Pests were unknown.

Today some of the substances mentioned above are no longer available. The combine harvester leaves chaff and weed seed on the field. Most farmers keep only one species of animals on pasture ground, but animals do not graze on areas where they have left droppings, therefore new ways to fertilize pastures have to be found.

To fertilize pastures we need well-rotted compost consisting of manure from different animals, and compost preparations. It is spread out when the Sun is in Virgo but at the latest in the last two weeks of October. Prior to fertilizing one can sprinkle small amounts of powdered basalt or wood-ash. At the following planting time one sprays Manure Concentrate Preparation in the afternoon to stimulate the processes of transformation.

Farmers keeping only one animal species can use Manure Concentrate preparations from other animals. The small amounts of manure required to produce Manure Concentrate Preparation can be obtained from neighbours, friends or gardeners. Manure of all domestic animals can be used. In the autumn they are stirred together and sprayed three times in the late afternoon, not later than the end of October, thus stimulating the micro-organisms, which multiply again in November and transform the soil. They decompose the dying roots, transform the rank patches and create a good soil structure. They also decompose leaves from trees and bushes.

In spring molehills are flattened and spread over the meadow and, in addition, Manure Concentrate is sprayed to activate the soil organisms. On the following Leaf days one can spray Horn Manure Preparation in the late afternoon to stimulate growth. If the plants have a height of about ten centimetres we spray Horn Silica for the first time early in the morning. If the farmer has enough time he can repeat it twice. Shortly before haymaking we recommend spraying Horn Silica on a Flower day to convert a usual strong protein process.

We made comparisons over many years, which ascertained that fodder dried in the sun yields the best long-term milk yield in the long run, and the healthiest animals.

Also, spraying the pastures with Horn Silica on Flower days is beneficial, as we have known for a long time that milk from protein-rich fodder, when consumed by babies causes diarrhoea.

Coming back to the hay, we have known for decades that the best hay quality is obtained when we mow during periods with Venus in front of the constellation Gemini. In addition, Flower days can be chosen. However, if there are planetary aspects in Taurus or Cancer and we mow on Leaf or Root days, the hay can go mouldy and cows do not take it with pleasure. Moreover it is bad for their digestion and supports microorganisms decomposing the milk. Often mouldy fodder causes lice. Similar effects occur with grass cut during the new Moon. For silage use Fruit days or alternative Flower days for mowing and silaging.

Root days are suitable for spraying Manure Concentrate Preparation on to mown meadows and grazed pastures towards evening. If new grass grows new roots are formed and the old ones die and have to be transformed by the soil organisms. The Manure Concentrate enhances this process, and spraying Horn Manure on the following Leaf days towards evening stimulates growth. When the grass has a height of about 10 to 15 centimetres it is time to spray Horn Silica on Leaf days in the morning, as mentioned already. In such a way meadows and pastures can be treated to obtain good growth and first class fodder.

Figure 92. Cow meadows at the farm of Wedig von Bonin.

Figure 93. Meadow's edge at Dexbach.

Harvesting

The harvesting of crops that are to be stored is best done on the same days that were favourable to the plant at sowing time. The exceptions to this rule are Leaf days. The crops we harvest on such days do not store well and soon decay. Instead of that one should use Root or Fruit days or, in the case of the cabbage family, Flower days. When harvesting fresh fruit one should use Fruit days within the time of the ascending Moon, while for root crops the Root days within planting time are the most suitable.

Times of harvesting

In general one can say that the times, which are favourable for sowing and cultivation of each kind of plant, are also good for harvesting. Leaf days are the exception. For fruits that are to be consumed immediately after harvest this is of lesser importance. In the case of fruits which are to be stored, or seed to be kept for future use, Leaf days must definitely be avoided since they easily encourage rotting or mildew on the seed. This is connected with the watery impulse on such days. One finds similar qualities after much watering; therefore plants grown under glass are generally unsuitable for winter storage. Even repeated watering of outdoor crops often causes them to rot when stored, whereas similar phenomena do not occur after repeated hoeing.

Harvesting trials with fruits

When gathering fresh fruit we noticed that the storage quality varied depending on the different constellations at harvesting time. If the fruit farmer wishes to take perishable fruit to market or to shops he will not want to offer the customers anything which will go off quickly. One summer we compared the harvest and storage of strawberries, cherries and gooseberries. At the same time we bottled fruit and made juice so that we could see how fruit from favourable and unfavourable harvest days behaved after processing. The juices from Leaf days soon developed mould. Besides this we made a sugarless jam from the fruits and found that the fruits from Flower and Fruit days kept best.

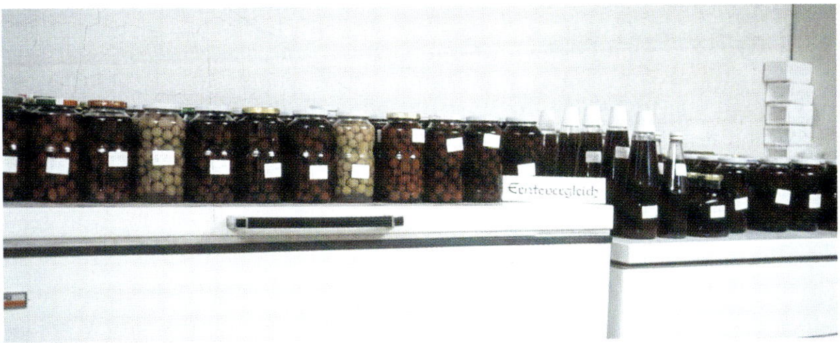

Figure 94. Trial of harvesting cherries (preserved).

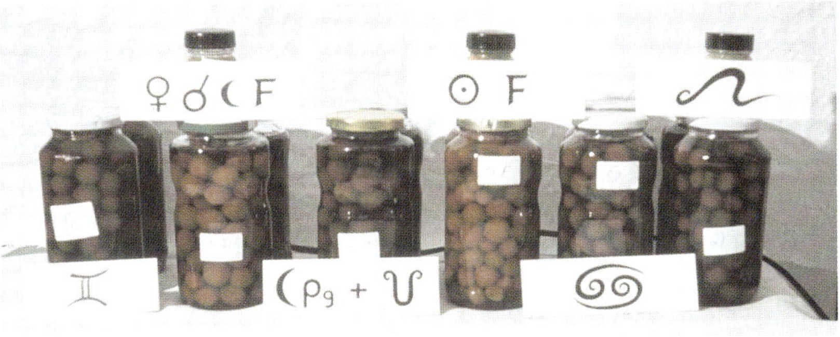

Figure 95. Cherries harvested at various aspects: Moon in Gemini, conjunction of Venus and Moon, eclipse of Venus, Moon at descending node and perigee, eclipse of Sun, Moon in Cancer, Moon in Leo (Left to right).

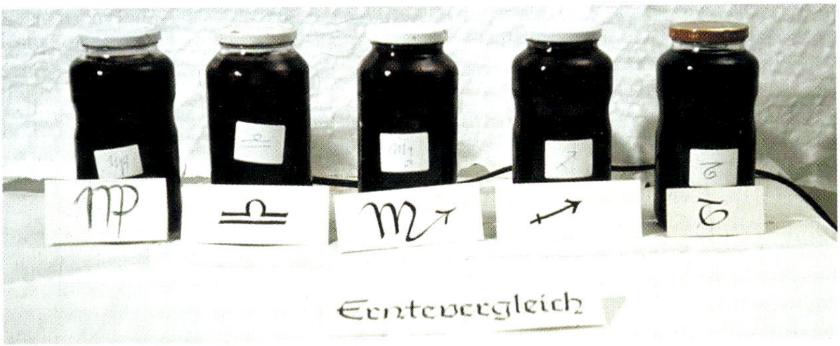

Figure 96. Cherries harvested at five different Moon aspects: Virgo, Libra, Scorpio, Sagittarius and Capricorn (Left to right).

Pea trials

When we compared results of using biodynamic sprays — especially the silica preparation — we discovered that the moment of application was of the greatest importance, since the preparation intensifies the effect of the cosmic influence on that day.

The yields could be increased when spraying took place on favourable days; during times with a negative cosmic aspect, the negative effect could also be intensified by means of the silica spray. When sprayed on Root days into the root region of the plant, it promoted the development of the fruiting part, but worked adversely on leaf and seed development.

Figure 97. Peas sown with the Moon in Sagittarius give a higher yield than when sown in Capricorn (Left to right).

Figure 98. Peas harvested during favourable and unfavourable aspects, and preserved to study the storage quality: Moon in Gemini, Moon in Cancer and eclipse of Venus, Moon at perigee and node, Moon in Leo and eclipse of Sun (Left to right).

Figure 99. Peas harvested with the Moon in Libra, Scorpio, Sagittarius and Capricorn to study the storage quality (Left to right).

Cereals

The amount of seed for sowing

Today it is recommended to use a large amount of seed and to leave small distances between rows. With a nitrogen-lime fertilizer the formation of strong ears is stimulated, but there is only one ear of each grain, which all grow to the same height. Only a few plants form a second or third stalk corresponding to a tillering factor of 1.7 to 1.8, i.e. that means not every plant has two ears. One can buy a thousand grains of seed which has a mean weight of about forty grams. In biodynamic farming we use less seed, as we like to achieve a good development of tillers.

Development of tillers

Cereals are monocotyledonous plants. In the past we observed the highest area density of ears and sometimes also the highest straw yield when newly-bought seed was sown on Flower days. However, the highest grain yield was obtained when sown on Fruit days (some visitors were interested in a high straw yield). But in the following year the neutral after-crops had the highest density of ears — and hence the highest tiller-factor — and the highest straw yield when sown on Fruit days, But newly-bought rye seed had the highest ear density when sown on Root days, and after-crops when sown on Fruit days. A commercially available oat seed had the highest straw yield when sown on Leaf days, in the following years, however, this changed to Fruit days.

Horn Silica was applied on Fruit days, after the plants had developed tillers, but not shot. Generally a higher density of ears was observed after spraying on Fruit days. The yield increase for sowing and spraying on Fruit days amounted to ten per cent for the grain and twenty-five per cent for the straw. Trials performed over many years revealed that the most important moment for the first spray of Horn Silica was at the two-leaf stage. So when the plant developed the second leaf, the first Horn Silica spray was made on the next Fruit day and the second about nine days later on a Fruit day, and the third one again about nine days later, also on a Fruit day.

Spraying Horn Silica on Fruit days at this early stage supports the development of tillers — i.e. more ears, longer ears and larger grain — finally resulting in a higher grain yield. This is valid for all cereals. In numerous trials growing after-crops we checked the reproductive forces and have found that cereals cultivated in this way had an excellent power of reproduction.

The following seven graphs show the grain weight of a thousand grains for seven different wheat species with the Moon in all twelve constellations and perigee. The highest weight and hence the highest yield was obtained for sowing and spraying Horn Silica on Fruit days (Aries, Leo and Sagittarius).

Figures 100–102. Cereals developing tillers, i.e. several plants grown from one grain.

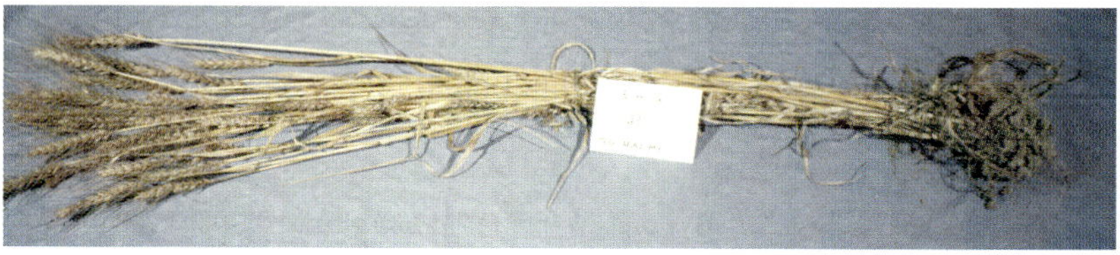

Figure 103. Ear-beds of various rye sowings.

Fig 104. Rye sown at various aspects.

Figure 105. Trials with rye.

Figure 106. Cereal trials at 'Thalacker.'

Figures 107 and 108. Cereal trials at 'Hell.'

Figure 109. (above left) Osiris-Pharao wheat.
Figure 110. (above) Megalith wheat.
Figure 111. (left) Wheat of M. Thun with awns.

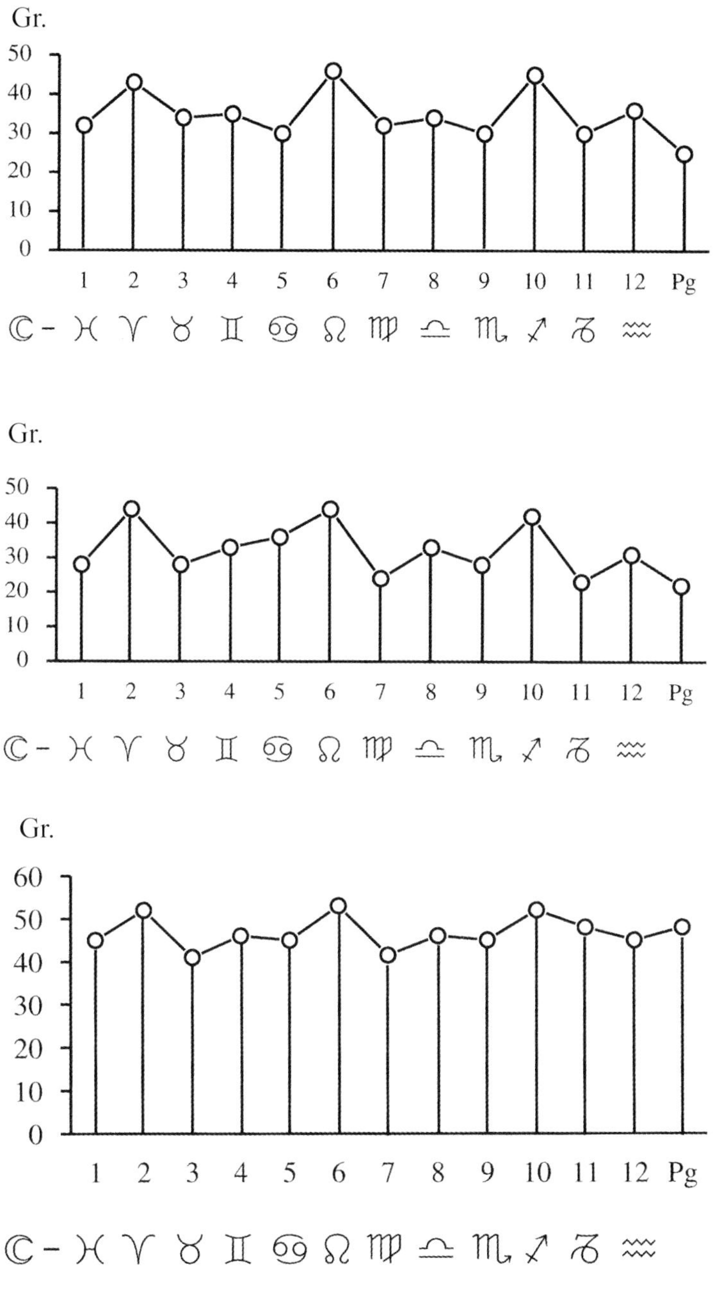

Tables of weight of 1000 grains. From top: Megalith, Osiris, Okasi.

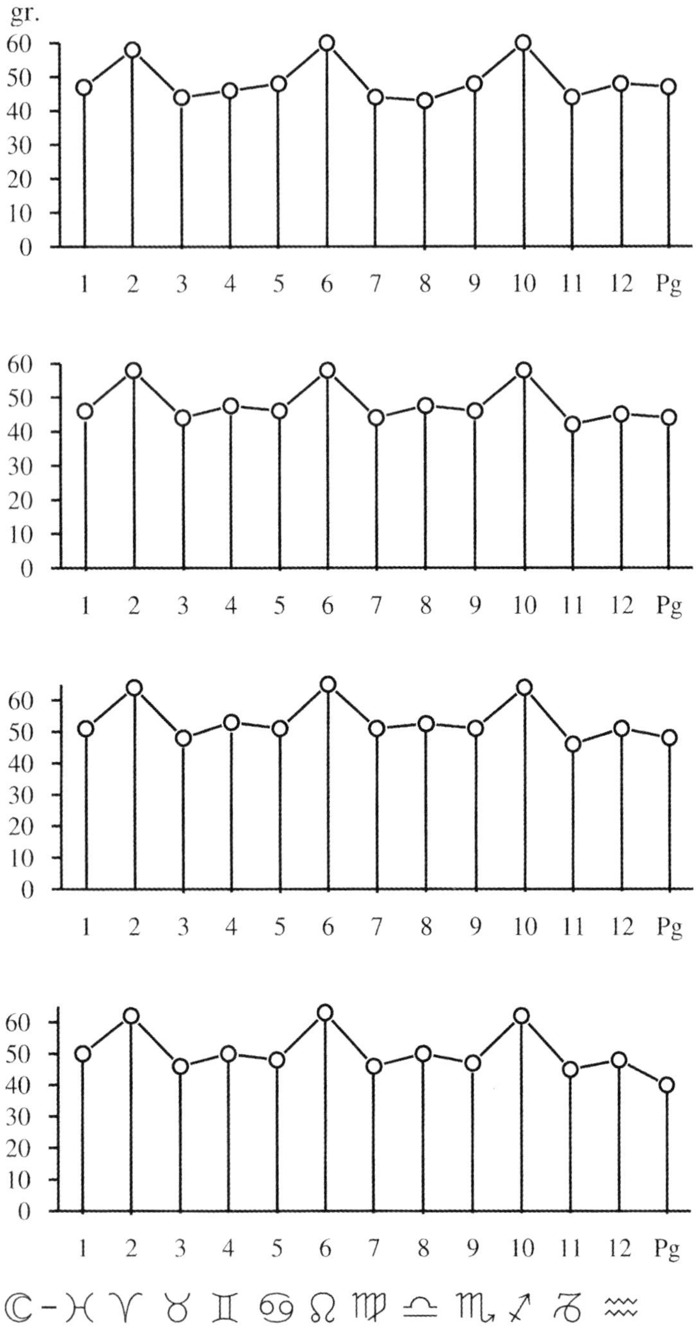

Tables of weight of 1000 grains. From top: blonde without awns; blonde with awns; brown without awns; brown with awns.

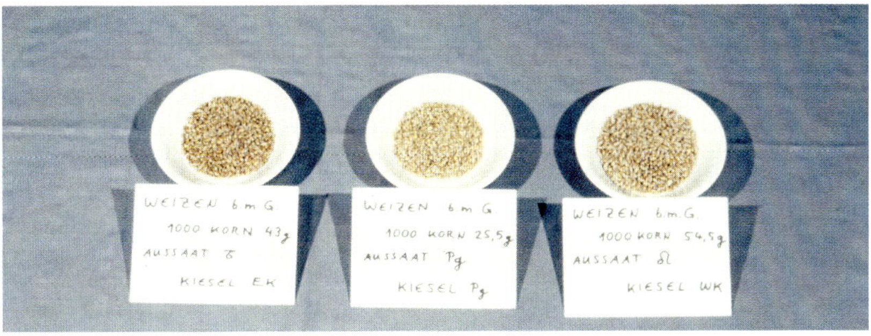

Figure 112. Thousand grains of wheat sown with Moon in Capricorn, at Perigee and in Leo (From left to right).

Processing of cereals

We carried out baking tests for all of our cereal trials, and so gained a lot of experience about the effects of various manures, aspects and preparations on the quality of the subsequent flour and bread.

We asked if the repetition of the same aspect can have a disadvantageous influence on the plant, and so we performed a series of sowings at the same aspect over many years. A wheat trial was performed over three years with sowings at fifteen different aspects. The results of three aspects were as follows. In Figure 112 each bowl contains one thousand grains. Left bowl: sown with Moon in Capricorn and three Horn Silica sprays on Root days. Bowl in the middle: sowing and three Horn Silica sprays done when the Moon was at perigee. Right bowl: sown with Moon in Leo and three Horn Silica sprays on Fruit days. The weight of one thousand grains is 43 grams for Capricorn, 25.5 grams for the Moon in perigee and 54.5 grams for Leo.

Growing Potatoes

In the past the recommendation for growing potatoes has been to manure in the spring with fresh manure for a fast-growing effect. However, by cultivating potatoes in all conceivable ways for almost fifty years, we have found that potatoes remain healthier if we manure the soil with 15 tonnes per hectare of well-rotted manure in the autumn. At first the manure is worked into the soil at planting time in October. At the same time Manure Concentrate Preparation is sprayed. About four weeks later, by mid-November, winter ploughing is done.

In November, micro-organisms have multiplied again and transform the soil. Ploughing later in December would disturb the activity of the soil organisms, the soil would not be transformed and there would also be a risk of substances washing out into the ground water.

In the spring, the soil is grubbered and harrowed. At each tending Horn Manure Preparation is sprayed. Potatoes are planted on Root days. As potatoes like the air it is best to hoe twice on Root days and spray Horn Silica at the same time. Together with hoeing and spraying the third time we hill up, so that the tubers do not become green.

In regions where *Phytophtora* occurs frequently, we recommend spraying stinging nettle three times in the late afternoon, on Leaf days. Put one kilogram of nettle leaves into five litres of water and bring it to the boil. After mixing with 100 litres water one can start to spray. It strengthens the leaves and the plants stay healthy.

Seed propagation

In extremely warm years potatoes can develop seeds. The fruits hang on the plants like tomatoes. Nearly seventy years ago the fruits turned yellow before the potatoes were harvested. Today one has to harvest the green fruits and let them ripen. They are then sown like tomatoes either in the greenhouse or in pots placed on the windowsill of a warm room. The best sowing time is on the last Root days when the Sun is still in Capricorn. At the two-leaf stage the plants are re-potted into peat pots in the same aspect. The soil is sprayed with Horn Manure and mixed well. In mid-May when they have reached a size of 12 to 15 centimetres, they are transplanted outdoors and treated as described above. The harvest occurs in autumn for seed for the following year. Among all aspects sowing when the Moon is in Capricorn, hoeing on Root days and harvesting with the Moon in Capricorn has been found to give the best results.

Propagation of eyes

To regenerate potatoes Rudolf Steiner recommended planting potatoes with middle-sized eyes. After many years of research we found that the best aspect is when the Moon is in Aries. Eyes located near the equator of the potato are cut out and are put three centimetres deep into the soil and ten centimetres apart. All tending and harvesting are best done on days with the Moon in Aries. The harvested potatoes are relatively small and can be planted and tended on Root days the following year.

Figure 113. Potato trial at Thalacker.

Regeneration of potatoes

To regenerate potatoes without special treatment good results can be obtained if one occasionally plants seed potatoes on days with the Moon in Aries. But one should only use tubers weighing less than 35 grams. The potatoes will not become very large but will produce the best seed for the next year. Hoeing and spraying Horn Silica should also be done when the Moon is in Aries.

Now we want to report on after-crops of potatoes grown from seed and planted, hoed and harvested during the same constellation for all twelve constellations of 1990. The tubers were planted, hoed and harvested with the Moon in all twelve constellations. In 1991, the seeds were sown for the first time (as described above) at the same aspects as in 1990. In 1992 the potatoes were planted at the same aspects as their seeds had been. Due to a lack of area we only planted at four aspects, on a Root day (Virgo), a Flower day (Libra), a Leaf day (Scorpio) and a Fruit day (Sagittarius), and sprayed Horn Silica three times at the same Moon aspects. In addition we had four control plots where no spray was applied. From these eight varieties we grew neutral after-crops in 1993 following the methods described above — 150 tubers of each variety were planted. The weight of the tubers was between 35 and 40 grams. They were planted on Root days and Horn Manure was sprayed three times. After germination we hoed and sprayed Horn Silica three times. The following yields were obtained:

Virgo	Root days	46.7 t/ha	49.2 t/ha
Libra	Flower days	44.1 t/ha	46.7 t/ha
Scorpio	Leaf days	42.8 t/ha	44.7 t/ha
Sagittarius	Fruit days	38.7 t/ha	39.5 t/ha

It should be emphasized that in 1993 optimal and uniform tending was done. The application of Horn Silica in 1992 improved the seed quality, and the propagation of potatoes via seed turned out to be the future method for biodynamic growers.

Cultivation of Oil-Bearing Fruits

As the livelihood of farmers has worsened over the years we have tried to grow less-known crop species to investigate how they react to biodynamic methods. For farm shop selling it is of great importance to have a diverse choice. Therefore, we started trials with oil-bearing fruits in 1989. The topic of the first trial was crop rotation. Contrary to what was expected we found that cereals grew well after oil-bearing crops. When we grew oil-bearing crops after Leguminosae, it was sufficient to spray Manure Concentrate Preparation three times in the autumn, and the Horn Manure Preparation three times in the spring in order to obtain a reasonable yield. Only winter rape behaved differently. To grow well it needed 10 tonnes per hectare of well-rotted manure applied several weeks before sowing, as well as the usual sprays.

Then we started sowing trials. Soon it turned out that the best seed yield was gained by sowing on Fruit days. Horn Silica was applied at the same aspect as sowing, and it became obvious that the oil content when spraying Horn Silica on Root days was less then for the control plot. This fact made hoeing and spraying trials necessary. Oil-bearing crops were sown in a large test field. After the plants came up they were divided into plots. On control plots no Horn Silica was sprayed.

There were four variations each with three plots. Every plot was sprayed three times early in the mornings:

AK	on	Leaf days
WK	on	Fruit days
EK	on	Root days
LK	on	Flower days

The yields of oil showed clearly: spraying on Leaf days yielded 5% more than without it (control plot); on Fruit days it was 8 to 10% more; on Flower days about 20% more, but on Root days about 4% less. The following oil yields (litres per hectare) are mean values over several years:

Species	Control plot	Leaf	Fruit	Root	Flower
Winter rape	380	399	418	365	463
Sunflower	358	376	394	344	430
Flax	360	378	396	346	428
Poppy	375	394	420	364	446
Oil radish*	324	340	353	308	389
Safflower	330	346	369	314	402

*Raphanus sativus oleiformis

Figure 114. Rape trial.
Figure 115. Rape oil: yields due to spraying Horn Silica at different aspects, right to left: Flower days, Root days, Fruit days, Leaf days, control plot.

Figure 116. Safflower trial at Thalacker.
Figure 117. Safflower oil: yields for spraying Horn Silica at different aspects, right to left: Flower days, Root days, Fruit days, Leaf days, control plot.

Figure 118. Sunflower trial.

Figure 119. Aime evaluating oil yields.

Figure 120. Sunflower oil: yields for spraying Horn Silica at different aspects, right to left: Flower days, Root days, Fruit days, Leaf days, control plot.

Figure 121. Flax trial.

Figure 122. Linseed oil: yields for spraying Horn Silica at different aspects, right to left: Flower days, Root days, Fruit days, Leaf days, control plot.

Oil-bearing fruits at Körtlinghausen

In the past rape has mainly been grown in Hesse/Germany, and occasionally flax to make linen. During the last few decades the cultivation of many different oil-bearing fruits in Europe has increased since species that ripen in colder regions became available. So we were often asked how our methods could be successfully applied to large-scale cultivation. The following example shows how.

Gyso von Bonin runs Körtlinghausen farm which has an area of 180 hectares located at the edge of Sauerland (Hesse)/Germany. He has organized the farm work in order to take advantage of the positive aspects and application of our biodynamic preparations. Oil is his main agricultural produce and he has a very good reputation. Taking account of his geographic and climatic situation he applies Horn Silica at an early stage of growth by spraying it in the autumn. He has observed that this helps the rape to come through the winter better. In the spring he sprays Horn Silica two more times. Flax is also tended carefully, taking the aspects into account. Another Demeter farmer located in a warmer region grows sunflowers for

Figure 123. Flax at 'Körtlinghausen'.

Figure 124. Oil mill at 'Körtlinghausen'.

Figure 125. Co-worker families in a rape field at the farm of Gyso von Bonin.

him. The oil of all three crops is mixed together and the oil yield amounts to 460 to 480 litres per hectare, and is of the best quality — it is 'shining'. His conventionally-farming neighbours attain neither the yield nor the quality.

On his farm Manure Concentrate Preparation and Horn Manure are also applied on favourable days whenever possible.

Trials with oil-bearing fruits

For some years we carried out similar experiments to those done with peas, but this time on oil-producing fruits. As far as growth and yields were concerned sowings on Fruit days proved to be the most favourable. From our comparisons of hoeing and spraying with the Horn Silica Preparation we discovered that the greatest yield of oil resulted when these operations were carried out on Flower days, in the later stages of growth. This was confirmed with sunflower, rapeseed, poppy and linseed. Since we have now acquired a press for rapeseed we can extend the experiments, repeat them more often, and make our own evaluations.

Figure 126. Oil-press of IBG Montforts, 41169 Mönchengladbach/ Germany.

Figure 127. Oil of linseed of an aspect trial.

Figure 128. Sunflower oil of an aspect trial.

Growing Brassicas

For 44 years we have performed brassica trials, and by comparing plant compost with manure from cow, pig, sheep, goat and horse, we have discovered that brassicas need well-rotted cow manure which has been treated with biodynamic compost preparations and worked into the soil in the autumn. Prior to sowings on Leaf days and to picking out, we spray Horn Manure or Hoof Manure. For transplantation to its final destination in the greenhouse, garden or field we use Leaf days during the planting time, spraying again Horn Manure beforehand.

When the plants have taken root and new leaves are growing we can begin to spray Horn Silica. We spray it three times, early in the morning on Leaf days, at intervals of about nine days. This leads to the formation of more substance and better aroma. If one cooks such a cauliflower there will be no fear that one can smell it throughout the whole house. Cabbage intended for storage or making sauerkraut (pickled cabbage) should be sprayed early in the morning with Horn Silica three weeks prior to harvest on Flower days. The best quality can

Figure 129. Cabbage trial in Dexbach.

Figure 130. Sauerkraut (pickled cabbage) trials.

be obtained by harvesting and cutting the cabbage on Flower days. Hoeing on Flower days some weeks before harvest has a similar effect. Herbal infusions can also be applied as mentioned in the section 'New methods to care for vines.' (see page 174).

Most species of cabbage, including cauliflower and kohlrabi grow best when sown and cultivated on Leaf days, with the exception of Broccoli, which grows best on Flower days. When the crops are well developed one should harvest as they start flowering otherwise.

White cabbage

In our courses about Manure Concentrate Preparation we recommend spraying it three times a year, but gardeners and farmers have applied it more frequently. Monsieur Sirlin, Alsace, France, grows white cabbage in order to produce sauerkraut (pickled cabbage), and complained about aphids and other animal pests.

Following our advice he applied well-rotted cow manure as well as the Manure Concentrate three times in autumn, but he also sprayed Manure Concentrate on the soil prior to planting white cabbage, and once a week during early growth. Naturally, he sprayed Horn Manure prior to planting, and Horn Silica later as recommended by us.

After one year he reported that the animal pests had disappeared and the cabbage was growing better, the yield had increased by 30%, and aroma and quality was better. During the winter he delivered 1000 kilograms of sauerkraut each week to the health food shops in Paris. He also reported that his cows in the cowshed had been biting one another but after spraying Manure Concentrate on the floor once a week the cows became very placid. Meanwhile Monsieur Sirlin died but his wife and son still enthusiastically apply the Preparations and for all work cosmic aspects are taken into account.

Figure 131. Monsieur Sirlin spraying white cabbage.

Figure 132. Huge barrels containing Sauerkraut of Demeter quality.

Biodynamic Wine Growing

Viticulture is as old as the cultivation of grain. After the Flood Noah became a wine farmer (Gen. 9:20). And after he had consumed enough of the fermented grape juice he was drunk. Fresh grape juice was not called wine. About 2000 years later scouts sent to Canaan by Moses came back with one grape that had to be carried by two persons (Num. 13:23). This was taken as sign of a very fertile land.

About 150 years ago we had the first collapse of wine growing. Vines had been imported from the American continent. But these had been grown from cuttings and not from seed. It is supposed that such plants did not have the same vitality. As a result the vine pest (*Phylloexera*) appeared and destroyed the vines in many European countries. Chemical industries developed pesticides and inorganic fertilizers that have been applied since then. The resistance of plants and the quality of juice and wine increasingly deteriorated. More pests appeared and as a result more pesticides had to be used.

A group of French wine growers, that had decided to work with the biodynamic method, asked us to support their efforts with the knowledge gained by our research on constellations. However, with the methods developed for farming and gardening, things changed very slowly. We realized that we had to activate the stubborn soil first. We had to find out the best compost for the soil. Which animal dung would have the best effect on these plants? Pig manure had to be excluded, or could only be added in small amounts to horse manure to make it less fiery. Cow manure has a positive effect mainly on the leaf region and should be applied sparingly. Following our comprehensive manure and compost trials we recommended applying totally humified sheep and goat manure to the vineyard at certain intervals. Poultry and pigeon manure can be added to cow manure. To all kinds of manure the biodynamic compost preparations are added (see 'The Manure Process', page 74).

Figure 133. After having sprayed manure concentrate working the land is done with horses.

Figure 134. Graf von Finckenstein mixing the substances of Manure Concentrate Preparation.

Working the soil in the vineyard

The manure spread in autumn is worked into the soil. Then the Manure Concentrate Preparation is sprayed in order to activate the micro-organisms and the transformation of the soil. During a period of conversion to the biodynamic method it is sprayed once a month on Root days in the late afternoon. Hoeing subsequently enhances its effect.

Vineyards and Horn Manure

Wine growers have frequently reported that there are fewer pests after the Manure Concentrate has been sprayed, but the biodynamic method comprises more preparations. Previously only their application was described, but now we are going to report on the production of the two Preparations. Rudolf Steiner speaks about horn and hoof of a cow. Calves do not have horns, they develop them later. The frontal bone grows and pushes the skin outside and this skin thickens to become the horn. In fact the horn develops by a growth process. The horn has tasks in the digestive system and this activity is still present for some time after the horns have been removed. This horn is filled with well-shaped cow dung and put into the soil over the winter. The horn's sucking ability attracts the Earth's winter forces and combines them with the cow dung, which brings about an intensification in the quality of the manure we use in a homeopathic way. The hoof — already formed at the embryonic stage of the cow — is also filled with cow dung and subsequently treated like horn. But it is important at what time we take the cow dung, and at what time we put the horn or hoof into the earth, or take it out. More about this question can be found in the section 'Horn and Hoof Manure Preparations' (see page 115).

The Horn and Hoof Manure Preparation has to be stirred for one hour. An amount of 140 grams stirred in 40 to 50 litres water is sufficient to spray an area of one metric hectare (about 2.5 acres). For a smaller area we use 30 grams stirred in 10 litres water which will spray about 2500 square metres (about 3000 square yards). A high-quality Preparation can be sprayed for up to 4 hours after it has been diluted. The Preparation will be effective for the whole of this time, and during these 4 hours the same area can be sprayed three times. For sowing and planting the effect is optimal when the soil is moved (by harrowing)

Figure 135. Manure Concentrate Preparation is ready for use.

Figure 136. Horn Manure and Hoof Manure.

Figure 137. Matthias stirring a large quantity in a wooden barrel.

Figure 138. David stirring a small quantity in clay pot.

Figure 139. Tractor with spraying device and barrel mounted in front and the harrow behind. This is the most effective and time-saving method.

between subsequent sprayings. For wine growers the best time to spray Horn or Hoof Manure is in March, prior to the beginning of vegetation. Both soil and vines have to be sprayed three times on Fruit days. If the sprays are made on one day over a few hours, we recommend moving the soil in between. As the spraying intensifies the cosmic aspect, unfavourable times in the calendar should be avoided. For large areas stir in large wooden barrels. The spraying device is mounted in front of the tractor and the harrow (or plough) behind it, so one can spray and move the soil at the same time. Small quantities are stirred and sprayed manually as shown in Figure 138.

Vineyards and Horn Silica

The chapter 'The Horn Silica Preparation' (page 111) contains a short description about the application of Horn Silica in cultivating plants. For its production the substances required are a cow horn and rock crystals. The rock crystals have to be broken up into small pieces and then ground with a pestle and mortar. They should not become a powder but still be fine-grained. Plants do not show any effect if they are as fine as flour. Then the fine-grained rock crystal is mixed with water and poured into a cow horn (we do not know if one can use a hoof since we have not yet made such experiments). On a Flower day near Ascension Day we put the horn about 20 centimetres deep into the Earth in such a way that no water can run into it. On a Flower day in the first half of November we take it out and put it into a glass in a bright place on the windowsill. Next spring it is ready for use. However, for vines the most important sprays have to be applied to the vine leaves after grape harvest. So depending on the climatic situation the Horn Silica Preparation can be used in late autumn when the leaves are still present. Horn Silica is always applied to the leaves so its properties can affect the cambium and return next year to the new leaves, vines and grapes. Ideally three sprays on Fruit days should be done early in the morning after sunrise. During summer no Horn Silica should be applied to the vine since the flavour of the grapes can be spoiled by the flavour of the leaves.

New methods to care for vines

In *Agriculture* Rudolf Steiner has pointed out that one should pay attention to how the plants that are necessary to make compost preparations act on human beings. For more than forty years we have occupied ourselves with these plants and their dietetic effect on human beings, animals and plants. For more than thirty-five years we have taken these plants and made teas which have been added to the bee food. They support the living forces of the bees and keep the bee colony healthy. Recipes can be found in our yearly calendar.

If we drink **yarrow flower** tea in the morning, we will wake up more quickly — this is especially effective on children, and if wine and fruit growers spray the same tea on the leaves on Flower days in the morning, the potassium process is activated and strengthens the plants in such a way that sulphurous sprays are not necessary.

The **camomile flower** tea has an antiphlogistic and antispasmodic effect and if we drink it in the evening we will have a pleasant sleep. Its strong relationship with carbon and formative forces can, for example, help prevent intestinal complaints such as diarrhoea. If wine and fruit growers spray the same tea on the leaves on the morning of Fruit days, it stimulates the calcium process and strengthens the plants so much that copper sprays are not required.

In *Agriculture* Rudolf Steiner speaks of the **stinging nettle** as a Jack-of-all-trades. In the section entitled 'The use of stinging nettle' (see page 130) applications have been mentioned already. Stinging nettle tea harmonizes all disorders between respiration and blood circulation in a wonderful way. It harmonizes lungs and bronchial tubes and stimulates water excretion. Spraying the same tea on vine leaves on the morning of leaf days stimulates the vital forces of the plants as well as iron and manganese processes.

Dandelion flower tea: Dandelion flowers can't be bought. We have to pick them from selected areas at a stage when the inner part is still closed and the dew is already dried. For harvesting Flower days mornings are most suitable. The tea strengthens the liver and harmonizes the processes between liver and skin, which can otherwise lead to skin allergies. Wine growers can spray the same tea on the leaves and vines on Flower days, early in the morning. It enhances the silica process in the leaves and prevents parasites from penetrating them. Above all this is of importance in rainy years, and years when the full Moon and perigee are occurring nearly at the same time.

Valerian flower tea: In medicine the root extract serves as a tranquillizer and sleeping drug. For bees we make the tea from dried valerian flowers to stimulate warmth processes. For plant growing either squeezed sap from flowers or tea from dried flowers can be used. On Flower days near Midsummer's Day, one can spray valerian on the soil

Figure 140. Nicolas Joly's vineyard near the mouth of the river Loire.

Figure 141. Nicolas Joly's vines.

Figure 142. Flowering valerian.

beneath the vines, fruit trees and beneath berry bushes in the late afternoon. It stimulates the bud development for the next year. After frosty nights we spray it on the affected plants. Because it activates a warmth process, the plants may wilt. To prevent this they need watering a few hours after spraying.

Tea production: In all cases fresh flowers can be used. If one takes dried flowers 10 grams are sufficient for a spray liquid of 100 litres. One can start with 5 litres and then dilute to 100 litres, and after stirring for several minutes it is ready for spraying. One hundred litres are sufficient for an area of two metric hectares (about 5 acres). With dried flowers of yarrow, camomile, dandelion and valerian an infusion is made with boiling water, which is strained after 15 minutes. Stinging nettle is put in cold water and brought to the boil, left to stand and strained after 10 minutes.

A French group of wine growers applied the above methods, and one can achieve good results by taking the guidelines seriously. One of them, Nicholas Joly, is putting his heart and soul into the work and uses his imagination to produce the best wine. The same is true for the processing of the grapes as reinforced by his clients and by experts.

Figure 143. The domaine of Monsieur François Bouchet.

The domaine of Monsieur Bouchet is located near Montreuil Bellay. From afar the plain looks reddish, like cherry trees in autumn with a green area in the middle. Getting closer we realized that all vineyards of conventional growing neighbours were suffering from an attack of mites *panonychus ulmi* (in German called red spider) while only a few of his plants on the edge were affected. Monsieur Bouchet has been growing wine biodynamically for longer than anyone else in France. His grapes and wines are judged to be so excellent that many wine growers want to switch over to his method. Now his son has taken over so Monsieur François Bouchet can advise interested wine growers.

Sundry Crops

Rape

Use Fruit days to sow, hoe and for the first Horn Silica spray to support seed formation. Further Silica sprays and harvest should be done on Flower days.

Clover

Leaf days are suitable for sowing clover for fodder and for the first Horn Silica spray. For seed production further spraying should be done on Fruit days. If the last growth is to be used for green manuring, spray Silica on Root days to enhance the formation of nitrogen in root nodules.

Red clover mowing trials

To obtain a weighed amount of food for our goat trials all plots were labelled, even the mown ones. Our colleagues realized that the growth was different depending on the day of mowing, so we performed similar trials over several years. We cut each plot twice and the third growth remained as green manure. Please note that the yields are related to a soil graded with 12 points. The red clover trial was done in 1988 and the following yields in metric tonnes per hectare (t/ha), were obtained by mowing twice:

Control plot	61 t/ha	100%
Leaf days	88 t/ha	143%
Fruit days	80 t/ha	130%
Root days	77 t/ha	125%
Flower days	76 t/ha	125%

After each cutting we sprayed Horn Silica twice at four different Moon aspects. However, during this trial, a final Horn Silica spray on a Flower day prior to mowing — to enhance the hay quality — was not made. The highest yield was obtained after spraying Horn Silica on Leaf days, and

Figure 144. Red clover mowing trial.

Figure 145. Berseem clover mowing trial at Gisselberg.

Figure 146. Berseem clover mowing trial at Dexbach .

was 43% higher than without spraying. Also for all other days a considerable yield increase was obtained. This clearly demonstrates that farmers can harvest food for more animals by applying Horn Silica.

Grains

Sow winter and summer cereals on Fruit days. Spray Horn Silica for the first time at the two-leaf stage on Fruit days and do a further two sprayings on following fruit days about nine days apart. Harvest only on suitable days.

Large-scale cultivation of carrots

Carrots like humus-rich soil without manure, which causes fast growing and grows favourably after rye. In early spring work the soil when the Moon is in Leo to stimulate the germination of weed seeds. Later work the soil and sow on Root days — preferably days when the Moon is in Capricorn which bring good crops and few weeds. When the plants have reached a height of about 10 centimetres spray Horn Silica for the first time on Root days and then at approximately nine-day intervals on Root days. Hoeing is also best done on Root days. Take care that the root collar is covered with soil to avoid chlorophyll formation. A green root collar spoils the taste and attracts pests. Spraying Horn Silica Preparation with the Moon in Aries or Libra, about four weeks before harvesting supports the ripening of the protein followed by a reduction in the nitrogen content and an increase in sugar content.

Field cultivation of beetroot

Beetroot grows best when the soil is fertilized with ten metric tonnes of mature cow compost per hectare. Beetroot can be sown either on Root days or on Leaf days. The first Horn Silica spray can be done after four weeks on Leaf days. A further two Horn Silica sprays and hoeing yield the best results when done on Root days. An application of Horn Silica in the autumn and harvesting are done in the same way as for carrots. To harvest avoid Leaf days and unfavourable times since they favour decomposition processes.

Strawberries

If we observe wild strawberries we can find that they flower in spring when leaves are poorly developed, then fruits grow and ripen. Later seeds develop and the fruits drop off. Only now the leaves start growing and at about the same time flower buds are formed for the next season. In autumn the leaves become red and fall off in winter.

If we are tending garden strawberries in accordance with the season we will have a rich harvest of aromatic and savoury fruit. This means that care should start after the harvest. To begin with we put the tendrils back into rows, and we put about two ten-litre buckets per ten square metres of well-rotted compost between rows. Be careful not to spread it onto the plants. The compost should be worked into the soil during planting on Fruit days. We also spray Manure Concentrate Preparation. The space between rows can be harrowed and sprayed with Horn Manure at the same time. To plant strawberries the same constellation is recommended and the best time is when Sun and Moon are in front of Leo, (only 3 days in the period from

Figure 147. Strawberry trials 1991: 2nd harvest (WK: Horn Silica spray on Fruit days, 0: control plot).

Figure 148. Flowering strawberries in the seventh year.

mid-August to mid-September, see *The Sowing & Planting Calendar*). Fertilize with compost and spray Manure Concentrate and Horn Manure Preparation. Then spray Horns Silica twice on Fruit days early in the morning and a third time also on a Fruit day but in the afternoon. The leaves will become red and this is the best condition for a good harvest next season. In the spring we only have to weed.

Strawberry trials

In September 1988 our strawberry field was planted. In the following two years, 1989 and 1990, hoeing, spraying and harvesting were done on days corresponding to the aspect of planting. However, in the third year (1991) the caring work and harvest was done on Fruit days in order to investigate the development of the plants under the influence of a rhythmic treatment for two years. A control plot was not sprayed with Horn Silica and was hoed under varying aspects. A total amount of 184.3 kilograms of aromatic and healthy fruits were harvested from an area of about 200 square metres.

The mean value of harvested fruits per plant:

Leaf days	632 grams
Fruit days	846 grams
Root days	636 grams
Flower days	666 grams
Control plot	460 grams
Edges	685 grams

We therefore recommend that cosmic fruit impulses are taken into account for growing strawberries.

Spinach seed

Three spinach plants are pictured below. In the middle a leafy type plant is shown. It was sown, hoed and sprayed on Leaf days, and weighed 600 grams and had fantastic quality. If spinach is not harvested it yields the best seed for the next season when hoed and sprayed with Horn Silica on Fruit days. The plant on the left-hand side is on the way to seed formation. The plant on the right-hand side represents a pure seed type with small leaves and yielded a large amount of seeds. These seeds if sown again don't develop many leaves but run to seed.

Figure 149. Development of spinach seed.

Calendula harvest trial

The teachers of the School for Horticulture at Hünibach (Switzerland) drew my attention several times to the fact that the day and time for cutting flowers had different effects on the plant. We had already done several harvest trials but for varied sowings. About thirty years ago we observed that cutting snapdragon, aster and zinnia on Flower days meant that the plants formed new flowers all the time. If flowers are used for medical purposes this becomes very important. In spring 1984 we sowed calendula in an area of 60 square metres using our own seeds. The total area was tended on Flower days. After flowering it was divided into plots and the flowers were picked at different cosmic aspects.

Picking 1st	2nd	3rd	4th	5th	total	
Leaf days	330	280	250	210	130	1200 grams
Fruit days	325	330	380	420	490	1945 grams
Root days	340	250	210	180	150	1130 grams
Flower days	335	380	510	590	650	2465 grams

Figure 150. Calendula trial.

Figure 151. Flower plants — aster.

It became evident that picking on Fruit days led to new flower formation but seeds soon developed. The same observation was made for cucumbers and beans. Picking on Root and Leaf days reduced the formation of flowers while on Flower days the flower formation was stimulated all the time without immediate seed formation. The plant was stimulated strongly to release pollen and scent.

Flower plants: Most flower plants do not like compost containing animal dung. When planting we apply horn manure and prior to flowering, and we spray horn silica one to three times on flower days early in the morning.

Herbs

Herbs react very sensitively to manure, and frequently animal dung spoils their good flavour. Pure plant compost is best when applied in autumn, with a subsequent spray of Manure Concentrate Preparation. In the spring we recommend spraying Horn Silica, as well as after each cut. Foliate herbs like Horn Silica on Leaf days but if we want to obtain a high content of essential oils we should apply it on Flower days.

Figure 152. Peppermint.

Figure 153. Lemon balm.

The Nature of Trees

The nature of trees is very different from annual plants. The rhythm of annual plants is dictated by the solar year: growth starts from the seed and is followed by the development of leaves, flowers and finally new seed. We can come closer to the forces of the nature by developing a feeling for the germination and shooting of the plants in spring or the flowering in summer. The scent and pollen seep into the macrocosm, and the macrocosm has a stimulating effect on the plants leading to a new seed formation.

In *Agriculture* Rudolf Steiner mentioned: "Each plant has its constellation such that a dandelion reproduces a dandelion and not a berberis." Both cosmic and Earth forces act together. By cultivating the soil we help plants to grow again every year. While wild growing plants die in the autumn and winter and life is only in the seeds, cultivated plants behave differently. By developing crops from the plant's root, leaf, stalk, flower and seed the plant gives up a part of its own nature to provide food for the human being.

Trees are different. They grow out of the soil having a strong upright force and in seed formation they cut themselves off from surrounding cosmic forces with their solid bark. The branched top forms a huge breathing surface and likes to 'breathe out' into the macrocosm during flowering. Remember the amazing scent of flowering lime trees. Then deciduous trees try to form seeds to lose both leaves and seeds in autumn, but life is not given up completely; it draws back inside the tree. Seed formation does not take place every year, only at certain intervals depending on planetary rhythms. Mars helps oaks, Jupiter, lime trees and Saturn, conifers. Moreover conifers change the needles at greater intervals.

Now we can ask what is common to all trees? Rudolf Steiner's descriptions in *Agriculture* are informative: "in the trees the solid, earthly element has in fact raised itself up into the air." He further describes living liquids, which circulate in the Earth and calls them 'Earth sap'. The Earth element with its life forces is obviously active in it. As it rises up into the tree it carries mineral substances, which are deposited in the wood. It loses its liveliness and becomes 'wood sap'.

In the leaves it encounters a life flow, which brings new forces from the periphery, is enlivened and becomes 'life sap'. The 'life sap' penetrates the tree enlivening the bark and the cambium. The combination of the warmth element and the cambium forms the basis for the effect of the starry forces being responsible for the tree's shape, and stimulates new fruiting.

Hence we find a fourfold life in the tree related to the four classic elements, which seems to stem from four streams originating from the Garden of Eden:

Earth element	Earth sap
Water element	wood sap
Air element	life sap
Warmth element	cambium

The four elements can be found again by regarding a slice of tree trunk. The outermost part is dead bark, followed by the inner bark and the phloem. These three covers protect the layer of fertility — the cambium — from which, in the case of deciduous trees, new 'annual plants' grow every year. They root in the cambium like annual plants in the soil.

Figure 154. Apple tree.

Figure 155. Apple tree (detail).

Thus the importance of caring for the bark becomes obvious. Fertilizing should be done via the bark together with tried and tested painting of the bark. If we grow plants like stinging nettle, calendula, phacelia and nasturtium below the tree the soil is harmonized and pests are kept away. A small amount of well-rotted compost given prior to planting enlivens the soil.

Horn Manure Preparation establishes a good relationship between the Earth sap and the wood sap. Together with warmth and light, applying Horn Silica at suitable times helps the trees to turn towards the periphery helping new starry forces to act in the cambium. By taking into account the cosmic rhythms for pruning and grafting we can stimulate special impulses from the joint effect of forces from the Earth and the periphery. In such a way we can perform creative work by caring for trees.

Sowing time for trees

Birch, lime, robinia, pear and larch are strongly influenced by the planets Venus and Mercury and can be sown and planted in the early part of the year.

Figure 156. Willow cuttings have become little trees in half a year.

Figure 157. Early growth of willow cuttings as fence.

Morello cherry, horse chestnut, walnut and oak are Mars trees and the ascending periods should be used (see *The Sowing & Planting Calendar*).

Maple, copper beech, sweet chestnut, apple and others have a connection with Jupiter forces.

For trees that are helped by Saturn there are good sowing times until 2004, i.e. for conifers, hornbeam, thuja, juniper, sloe and plum.

In all cases, one should choose the days which are experiencing an opposition of the corresponding planet, within the suitable times if possible.

Willows

In Figures 156 and 157 willow cuttings are shown which were put into the soil in March, and by autumn they had roots and branches and could be transplanted as little trees. In the same way one can grow hedges and hedgerows as fences by simply putting willow cuttings cross-ways thirty centimetres deep into the soil during planting time in March.

Figure 158. Christmas tree with sprouts.

Cutting Christmas trees

When planting time ends the Moon becomes ascending and this period is recommended for cutting Christmas trees which then stay fresh for a long time, have a delicate scent and sometimes sprout.

Painting of fruit trees

To treat the stems of fruit trees take cow dung and clay in equal quantities and stir with whey until the mixture has roughly the consistency of whitewash. Then clean the trunks and thicker branches with a wire brush and paint them. To paint with a syringe it is necessary to dilute the paint with water.

Manure Concentrate Preparation is most effective when sprayed on the trunk and branches, and also on the soil under the tree on a frost-free winter's day. Otherwise it can be added to the mixture for painting. Horn Manure Preparation should be sprayed under the trees in spring, and Horn Silica should be applied three times on Fruit days after the harvest.

Cutting scions for grafting

Just as the plant's root system is more active during the descending Moon, so the plant lives more on the periphery during the period of the ascending Moon. The forces and saps in the plant tend to rise more into the upper part, into the periphery. If fruit is harvested at this time it can be stored for much longer. This time is not suitable for cutting timber. If one is taking scions during this time for grafting on to a host plant then one should choose Fruit days as the plants respond with healthy growth and better yields.

The Forest

In a healthy natural organism — or an ecosystem as it is called today — the forest has always played the greatest imaginable part. It is of decisive importance in connection with the regeneration of agriculture as well as healing influence.

With the development of the industrial society, forests that had been there for thousands of years began to be felled. On all continents large areas of forest disappeared. Only in rare cases were they replaced by new stands of mixed forest. Even in Europe it was thought that time was too short to allow slow-growing trees to be replanted, so conifers were mostly planted, which guaranteed quick growth. Cosmic rhythms were no longer taken into consideration.

In his agriculture course Rudolf Steiner mentions that trees can be seen in connection with planetary rhythms. This course forms the basis for biodynamic agriculture and was given in 1924. He said to the farmers that the forester who plants oak trees should be familiar with the Mars rhythms, and people planting conifers should know about Saturn rhythms.

> ... if you wish to plant coniferous forests, where Saturn forces play so great a part, the result will be different if you plant the forest in a so called *ascending* period of Saturn, or in some other Saturn period. One who understands can tell precisely, from the things that will grow or will not grow, whether or not they have been planted with an understanding of the connections of these forces.

Many years ago Kurt Willmannas, a forestry expert reported on records made of a large area of forest in Hesse, Germany over a period of 130 years. These showed that those tracts which were planted at favourable planting times were still healthy whereas other areas, planted at times of unfavourable rhythms, had long since been attacked by the bark beetle and other pests.

If we look nowadays at the problem of European forests — referred to as “dying forests” — then we can see that many factors (as described by the media) are linked with environmental pollution. The picture of the damage has many sides to it. Conifers no longer have so many annual rings as before and so there is less growth of wood. This was first shown when felling trees. Then there is another phenomenon. A kind of crippling condition is appearing. If the capillaries are open at the moment sugars are formed then they remain open and the tree's sap evaporates in a short time. Sugars are then formed which encourage attack by weevils and other insects. One tended to ascribe the dying of the tree to insect attack and took steps to combat it but there was no longer anything to save. If this condition occurs when the capillaries are closed then they never open again. The tree is unable to exhale the rising sap and drowns in its own juice. On cutting down such trees one finds that the lower part of the trunk is filled with sap. To reduce the collection of sap fungi appear but they are not the cause of the problem. Combating these fungi is no use and the tree still dies.

If one looks for the time these deaths were first noticed among certain species of trees then one finds the first reports around 1860 in Silesia (now Poland). In about 1890 there were reports of dying firs about 150 km further to the west, and in 1920 a new period began in Saxony and the Vogtland. By 1950 a new wave of dying forests occurred, this time in the Sudeten and the Erzgebirge. In the meantime whole areas have died off. A new factor emerges here. In the mountain valleys even grain, potatoes and vegetables do not grow any more. The last wave began in the Bavarian Forest in 1975/76, and continued northward along mountain chains. When the 30-year rhythm occurred in 1980, "dying forests" had spread all over Europe.

During earlier periods the dying came to a standstill after a few years. This is not the case now. The whole of the environment is so damaged in its life forces, the fertility of the soil and its ecological wholeness (through overdoses of chemicals, radioactivity, radar waves, pollutants from power stations and motor vehicles) that the consequences are incalculable. The symbiotic relationships and micro-organisms which every plant needs in its root system for its whole growth have been destroyed in many areas and the root hairs have begun to wither. Mineral fertilizers do not help here any more.

If we do not succeed in finding the true being of a tree and the soil then fertile land will become desert. There are already many existing deserts on our Earth, which were started by human beings using the wrong methods. They are growing larger all the time.

Rudolf Steiner characterized trees to the farmers as "heaped-up

Earth." He spoke of living liquids, which circulated within the Earth and called them "Earth sap."* The Earth element with its life forces is obviously active. As it rises up into the tree it becomes more of a chemical substance and allows the minerals to die into the wood. We can sense the element of water active in dissolving and condensing. The watery part moves upwards into the leaves, where it meets the airy and light forces in the periphery, and is then newly activated. In this assimilation the elements of air and light are active. In the relationship with the periphery starry forces are absorbed and these are taken up through a warmth process into the cambium. Year after year this provides the cosmic nourishment for new growth on the wood.

There is a fourfold connection, which can be summarized as follows:

Earth sap	the element of Earth
wood sap	the element of water
life sap	air and light
cambium	the element of warmth

We can see in this the order of the traditional elements, which we have reported on during the past years of experiments: they form the main link in mediating cosmic rhythms and forces. When caring for trees this fourfold order should be taken into consideration. From all this arises the need to apply a manure to the tree bark. We are referring to the mixture of clay, cow dung and whey — the healing properties of which are already widely recognized — which is painted on the bark. At the beginning of this century there were health spas, which worked entirely with whey treatments. In addition to the aforementioned ingredients for painting on the trunk or spraying on to the upper branches, we can add some wood ash and some ground basalt. After two or three treatments the bark is already totally renewed and with that one achieves rejuvenation in the cambium formation as well. In the root region must pay attention to the right development of the soil organisms, but not by applying large amounts of manure. This is not only applicable for forest trees but also for fruit trees and bushes.

Let us return to the question of cosmic rhythms. It is said that 120-year-old fir trees, 90-year-old spruce trees, 60-year-old pine trees are

* Rudolf Steiner, *Mensch und Welt*. GA 351. Lecture of Oct 31, 1923.

dying. With Saturn we find 30-year rhythms due to the length of its orbit that means that certain effects are repeated every 30 years. Let us look at the planetary nodes. Those are the times when the Moon or the planets intersect the ecliptic. Owing to the differing positions of the planets' paths from the earthly point of view there arise points of intersection, which then give rise to eclipses. In many series of experiments we discovered that plants sown or planted around such nodal points revealed disturbances in growth, which then continued on into the following generations through the seeds. In August 1975 Saturn was in such a nodal position, and its negative effect stretched over about 3°. It took months for Saturn to move on slowly since the retrograde movement also fell in this time and the planet was in its descending period. However there is also another factor.

The planets mostly have their own moons, which encircle them just as we have "our" Moon on the Earth. Saturn has twelve moons, which encircle it at different speeds. In addition to this a ring surrounds it. Apart from Saturn's own rotation there is the rotation of this ring and a shift in the position of the ring. Sometimes its uppermost side faces the Earth; sometimes it is the lower side. Halfway between the descending and ascending periods it moves as if through a kind of neutral position (like going through the eye of a needle). The ring becomes invisible and exists only as an invisible line. The question that arises is: does it block the influence of Saturn on the Earth at this time?

We may be sure of one thing at least. The moment we have described above, which falls in the descending Saturn rhythm, is in keeping with the dates from forestry information: 1860/61 – 1890/91 – 1920/21 – 1950/51 – 1980/81, and are times in which the life of the trees was seriously threatened. According to our research into cosmic rhythms there are no chance events and we should take this very seriously in view of the difficulties in the environment. In the course of the last few years we have found that something more must be added to this. In our Planting Calendars we have always pointed to the fact that during recent years all the planets have been standing together for some months in close cosmic proximity. Therefore many conjunctions took place when the planets near the Earth screened off or weakened the influences from the outer planets. This was the case in December 1983 and January/February 1984. From June 1984 there were no oppositions, then all the outer planets were affected again by conjunctions. There have been months during the past few years when the forces which are

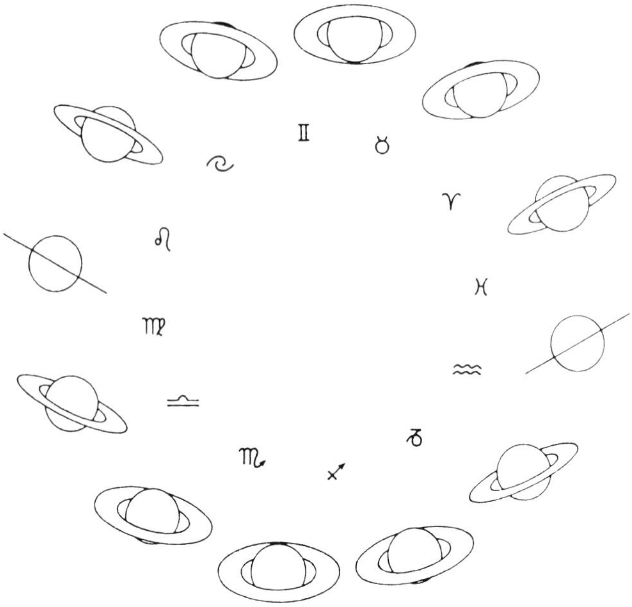

Figure 159. The changing position of Saturn's ring as seen from the Earth during one revolution through the zodiac. Saturn begins to ascend at 270° and to descend at 90°. In 1975 the ascending node was at 116° in Gemini and in 1990 the descending one was at 289° in Sagittarius. From Movement and Rhythms of the Stars *by Joachim Schultz (Floris Books).*

not only important for conifers — but also for beeches, oaks, sycamores and other trees — have been very weak in their influence. It appears high time that people direct their attention to the question of cosmic rhythms in forestry, which begins at the time one harvests the seeds, sows them, plants them out and carries out cultivations.

As we can no longer reckon with natural soil fertility, in view of environmental pollution caused by technology and the use of chemicals, we are called upon to treat the soil and plants in a way that can still salvage something from the Earth on the basis of real spiritual insight. In recent times it has always been stressed that something can be proved by scientific methods. However when looking at the situation of the forests we have to admit that we are ignorant and that Rudolf Steiner was one of the few 'in the know.' To answer the questions that have been raised, he has

Figure 160 and 161. Trying our new syringe to spray fields and woods.

bequeathed the biodynamic preparations to sustain and regenerate the fertility of the soil. His recommendation to various farmers who took part in his agricultural course was: "Apply these compost preparations in as many places as possible so that some green oases remain in Europe."

Following many experiments with compost preparations it can be seen that they encourage the breakdown and transformation of manure as well as keeping substance loss to a minimum. They bring about an increase in micro-organisms in the compost and the soil and promote a good clay-humus complex, good soil structure and increased quality in the soil. All this has disappeared from soils where the forest is dying. Through an increase in environmental pollutants

Figure 162 and 163. Spraying in the woods.

one increasingly finds that unfavourable cosmic constellations have an influence.

One may also add the following to Rudolf Steiner's remarks from his agricultural course: "This Saturn ring is something quite different from what the astronomers say about it: this ring is circulating health and the inner part of Saturn is sickness, seen in its purest concentrated form." A further question leads on from this. Do the healing powers of the ring, when invisible during Saturn's descending period, get diverted into other regions of the cosmos, but not towards the Earth?

When speaking of the dying forests we pointed to the enlivening effect of biodynamic preparations. With the first experiments we carried out in forested areas we observed a slightly more powerful green in the treated areas from an aeroplane. Unfortunately spraying from aircraft could not be repeated often enough due to its cost. In the meantime a spray has been found that not only works on fields but also on forests. In this way by means of the manure concentrate, the compost preparations can be sprayed more often into the woods. Following this one can then apply Preparations 500 and 501.

We hope that these are the first positive steps in answering the problem of the forests. We thank all those who, sparing neither effort nor expense, enabled us to put these and other recommendations to practical use.

PART FOUR

Pests, Diseases and other problems

Weed Problems

When we were carrying out cultivations during our experiments with constellations we noticed that weeds around the sown areas did not grow uniformly, due to the different influences of the constellations. In 1968 we began experiments with weeds, which were more or less completed by 1979, leaving only a few isolated questions about treatments of large areas and the methods one could use there. These needed further research. As a separate publication about our experiments and results with weeds has already appeared (in German) we shall only point here to a few of our findings. With our work on the soil divided between different kinds of plants, we first discovered that weeds reacted in different ways at the stage of germination, according to the Moon-zodiac influences, as well as planetary aspects. Work on the soil during Lion days (when the Moon is in the constellation of Leo) stimulates the germination of many more weeds whereas work on Goat days produces only a few weeds.

Following an indication by Rudolf Steiner to burn the seeds of weeds and scatter their ashes over the fields to repress the weeds, we carried out experiments over twelve years. We were able to confirm that the effects worked over periods of four years. We noted an increase in the effects when the ashes had been stirred for an hour prior to their application.

We looked at the question of homeopathic potencies and their effects when used over large areas. In ten years of experiments we tested their methods of production, their application and their effects. During the last few years of this work we devoted our energies to the questions:

1) When applying these potentized ashes should one bear in mind the constellations?

2) Is it possible to influence different weeds with the same application?

Do these weed ashes or their potentized forms affect cultivated plants? The results show new possibilities of dealing with weeds that will certainly be of importance for farming and gardening. To describe them further would go beyond the limits of this book since it is a specialized area. We trust the reader will appreciate this.

Plant Diseases

We speak of plant diseases when cultivated plants are affected by plant parasites. The wrong type of manure, or manure that has not sufficiently matured, often causes these. Generally speaking, seeds bought nowadays lack genuine life forces. Fungi attack them almost immediately they touch the soil; after all it is the function of fungi to cause dying organic substances to decompose. Seed dressings of certain types can avoid early damage to the seed. A second fungal attack often occurs when the plant is just past its earliest growth period. Prophylactic sprays are used, therefore, in the early stages.

However, preventative treatment should really be of a different kind. The first pre-condition is the good regenerative power of the seed itself. If sown into a well-matured soil the young plant can develop without the prior help of seed-dressings, but if processes of fungal decomposition are still in full swing this fungal activity can be carried upward into the plant. How do we then bring it back to the soil again? One way of doing this is to hoe frequently in the evening. By stimulating activity in the soil the level of fungal life recedes. In severe cases an application of horsetail (*Equisetum arvense*) will help if one takes 10 grams ($^1/_3$ oz) of the dried herb and brings it to the boil in 10 litres ($2^1/_4$ gallons) of water and, when cooled to about blood-heat, sprays the plants and soil with it, again in the evenings. It is advisable to spray the soil the next morning with diluted stinging nettle manure (see page 130). The effect of Equisetum is to contain fungal activity in the region of the soil, while stinging nettle stimulates renewed and healthy growth.

If Full Moon and the Moon's Perigee coincide on the same day an increase in fungal attacks can be expected during these years. When it occurs attention will be drawn to these critical times in the *Sowing and Planting Calendar*.

Fungal activity is also strongly affected by the time of harvesting as far as subsequent sowings are concerned. We found, for example, that if harvest had taken place on a Leaf Day, at Perigee, at a Water Trine, or on the day of a node or eclipse, corn sown the next year tended to be very susceptible to fungal attack.

Animal Pests

If we succeed in becoming familiar with the ways of so-called animal pests then mistakes we have made will often in turn make it possible to regulate things merely by changing our working methods a little.

The turnip gall weevil, *Tohynchus pleurostigma,* a small brownish-black beetle, will assail plantings of late cabbage when both Sun and Moon are in front of Taurus, especially if young plants are still standing close, perhaps too close, together, so the individual plant has not enough space around it. The beetle pierces the neck of the root and lays its egg in it. The plant then forms a kind of sheath around the growing embryo so that it can develop. If we open such a sheath — sometimes also called a gall — then we find a small larva lying there that is an in-between stage in the development of the beetle. It perishes when air gets to it. Later when the plants have been planted in their final position and have sufficient space around them the beetle shows no interest in them any more.

One can, of course, cover up the seed-boxes or beds during the above-mentioned three days so that the beetle cannot get to the cabbage plants.

It is different with the cabbage root fly (*Delia brassica*). It also lays its egg in the neck of the root. The larvae then eat away the tender cambium layer of the lower stem and upper root so that the plant dies off. The cabbage root fly attacks and lays its eggs when Sun and Moon stand in front of Aries; early cabbages are affected in this case. It is not the young plants in seed-boxes, which are endangered now, but those, which have had their final transplanting but were planted too deep into the soil. A small section of the stem, which grew in the sunlight before has now been brought into the darkness of the Earth and there provides a suitable basis for the early development of the cabbage root fly. If the attack is noticed early enough the leaves go limp in the middle of the day, a twice repeated treatment with wormwood tea may save the crops: 10 grams ($^1/_3$ oz) of the dried herb are brought to the boil in 10 litres ($2^1/_4$ gallons) of water, and each plant is given a dessertspoonful directly on the stem. The same method could also be applied in the case of other so-called animal pests.

However, if a direct attack has to be controlled, Hahnemann's principle of homeopathy, i.e. of treating like with like, can be applied. Rudolf Steiner suggested rotting the offending insects in water, or burning them. In either case the remains of rotting or burning should be scattered in infested areas. Some examples follow.

If mole crickets, *Gryllotalpa vulgaris*, (hardly known in Britain) are a nuisance, one should burn some of them in a wood fire when the Moon is in Scorpio and the Sun in front of Taurus, then rub the ash to a fine powder in a mortar or bowl and then apply it to their tunnels. We have had good results with the scattered ashes of Colorado beetle (a notifiable pest) and their larvae, the burning having been carried out with both the Sun and the Moon in front of Taurus and the ashes rubbed down for an hour. The ashes can also be kept and spread in the autumn or in spring on any areas where potatoes are to be planted. If a homeopathic decimal potency of D8 is made of the well-powdered ashes and sprayed three times within two days the effect will be enhanced.

If mice or birds become a pest some skins (with their feathers on in the case of birds) should be burnt when Venus is in Scorpio and the Moon is in Taurus, and the ashes are then scattered in the respective places.

Homeopathic methods of regulating pests

When animal pests appear one should at first try to get an impression of their life cycles then one can often discover the mistakes made which led to the attack. In this way one can intervene and regulate them. In many cases direct intervention is necessary. Homeopathic methods can be recommended, that many users have reported as having astonishing effects in some cases, if our recommendations are carried out very precisely.

It is best to take fifty or sixty of the relevant pests and to burn them in a stove with a good fire during the appropriate constellation. The ashes, which are left, should be pulverised in a mortar for one hour — i.e. 'dynamized' — then take one gram of this ash mixture, put it in a bottle with 9 grams of water, and shake it for three minutes. Now one has the first decimal potency. Add 90 grams of water and shake again for three minutes. Now one has D2. When one proceeds in this way one reaches D8 with 100,000 litres of water. Naturally one can't cope with this. Therefore one goes up to D4 and then begins again with smaller amounts. When one applies this process, certain deterrent effects appear at D8, if one has applied it in a fine spray on three evenings in succession. We have received reports of good successes with different creatures:

stable fly	Sun and Moon in Gemini
wax moth	Sun and Moon in Aries
cabbage-fly	Sun and Moon in Taurus
Colorado beetle and Varroa mite	likewise Sun and Moon in Taurus
cricket	Sun in Taurus and Moon in Scorpio

After about four weeks one can start again with the same Moon aspect using D4, lead up to D8 step by step, and then use these sprays three times. When applying prophylactic sprays against Varroa mite cover an area of about 15 metres around the beehive. If the pest attacks directly spray in the flight exits and between the combs.

We can also use the same methods against slugs and snails by taking advantage of Moon in Cancer — three sprays of D8 on the ground of the whole garden. One can also apply the residue of what was burnt, but then one needs more insects.

Using these methods one contains the different groups within their limits and prevents uncontrolled reproduction. Many species undertake such regulation themselves, as one knows from the example of lemmings, which during certain Venus constellations every eight years, rush to the sea and drown themselves.

PART FIVE

Food Preparation

Bread

Recipe for rye bread

Since all the grains we grow in our experiments were tested for quality, we have used our own raising agents (without additives) for bread baking to produce a good loaf. Besides sour milk, buttermilk, whey and syrup we have also tried to bake with honey and have developed a recipe that has proved its worth over many years. One heaped teaspoonful of flower honey is stirred well in a glass of warm water (50°C) and then mixed with $^1/_4$ lb (250g) of finely-ground rye meal. This small amount of dough is made in the evening and kept warm overnight. It should be at a temperature of about 26–30°C by the stove or next to a hot plate, which is set very low. Next morning one adds the same amount of rye meal and warm water or whey. In the evening add the rest of the flour (approximately 60% of the total) to the prepared

Figure 164. Rye trials.

Figure 165. Breads of the rye trials: two examples, rye plus Horn Silica (501) and rye (from the right).

dough with sufficient warm water. At this stage one can add a little linseed, caraway, fennel or something similar and leave it to rise overnight. Next morning one adds salt and finishes the dough. When it begins to rise again the loaves are formed. Let them rise well, put them in a preheated oven, and bake them for a good hour. Rye is easy to digest when it has gone through these five stages. One can keep about 500 grams of the finished dough and leave it in an earthenware pot. After it has risen again a little, it should be sprinkled with salt, covered with greaseproof paper and stored in a cool place such as the cellar (not the fridge).

When one wants to do some more baking take the pot out of the cellar in the morning and add a teaspoon of honey, which has been stirred in a glass of warm water. Then keep the pot warm. In the evening one can start on the main dough and proceed as described above. One can also begin from stage one again but then it will take longer with this kind of sour dough. Rye should rise five times. Wheat, barley and oats need to rise only three steps. Success depends on the warmth of the baking area.

Making bread from rye, wheat, barley and oats

In the last twenty years Europeans have wanted a larger variety of bread and so today bakeries offer numerous kinds. One can speak of a distinct bread culture being honoured by the consumer. However, people on farms and in their homes who make their own bread may have fewer possibilities. Therefore, we developed a method for making different kinds of bread at the same time and without extra effort.

The recipe

Take 250 grams of wheat, rye, barley and oats and put them into four bowls, together with half a litre of cold water and leave to soak overnight. If in a hurry one can put the bowls close to a warm stove or put them on the four hotplates of an electric cooker at the lowest setting, to warm the water up slowly — but one has to avoid boiling. After about two hours the water should have been absorbed.

In the meantime mix 1250 grams wholemeal (wheat) with 1.5 litres hand-hot water to which a bit of yeast has been added. Take 7 grams of dried yeast, add two teaspoonfuls of sugar and stir in a cup of lukewarm water. Wait for ten minutes before adding to the wholemeal dough.

If the grains in the four bowls have soaked up the water allow them to cool. When the temperature is hand-hot add 250 grams dough to each bowl, plus one teaspoonful of salt. Add a teaspoonful

Figure 166. The wholemeal dough is well risen (top right), barley, rye (2nd line from left) and oats, wheat (last line from the left).
Figure 167. Grains soaked overnight, barley and rye (top from left to right), oats and wheat.

Figure 168. Four breads: rye, wheat, barley and oats (from top).

of caraway to the rye, two teaspoonfuls of linseed to the wheat, one teaspoonful of aniseed to the barley and one teaspoonful of fennel to the oats. When using whole seed spices we recommend adding them at the beginning to the cold water together with the grains. Add the ground spices now, knead the dough well and let it rise. This takes about half an hour.

We like the spices given above but personal favourites can be used as well as any other dough suitable for bread-making.

When the dough has risen knead with another 50 grams flour and put in a baking tin which has been greased or covered with coarsely ground wheat. Leave the dough to rise again for about 20 minutes, then put into a preheated oven at 150°C. After about 10 minutes increase the temperature to 200°C. After another 50 minutes the bread should be ready, but it is better to check, as it may vary by five minutes or so.

To the last part of the wholemeal dough add a teaspoonful of salt, 100 grams of raisins or dried fruit. Knead the dough together with some flour until it is dry and put it into a baking tin. Let it rise for half an hour and put it into a preheated oven at 180°C. After about 45 to 50 minutes our Sunday bread is ready. Finally, we have five different breads of different nutritional quality.

Figure 169. Pure rye bread. Figure 170. Our Sunday bread: wheat and raisins.

Here we mainly eat rye bread made from whole grain since this was the traditional regional bread. Readers frequently ask how much flour can be added to the wholemeal dough described above. Actually it can be used for any amount of flour but the more one uses the longer it needs to rise.

To obtain good bread the room has to be warm enough — that is at least 28°C for rye — because at lower temperatures the bread becomes sour. The lactic acid bacteria need this temperature to multiply. If the temperature is too low acetobacter develops making the bread sour and indigestible.

Recently Finnish scientists have found that each human being has cancer cells in his organism that are expelled after eating rye bread. At the university of Giessen (Germany) Prof. Wagner implanted cancer cells into rats and studied their development as a consequence of drinking milk from different animals. The cancer cells atrophied after goat's milk and to a lesser extent after sheep's milk. Although it is ambiguous to transfer such results on to people, we have evidence that goat's and sheep's milk can be a helpful part of a diet for people at risk. This means that well-chosen food can have a curative effect. We try hard at our research station to develop recommendations for the cultivation of crops which are best for people, animals, plants and the Earth.

Figure 171. Performing baking tests with grain from a manure trial.

Figure 172. Baking tests: Bread of wheat taken from manure trials.

Baking tests with wheat

With a great variety of fruit, the times of sowing and harvesting, the manuring and the cultivation methods show up not only in the quality of seed and storage, but also in the processing. This became very clear with our experiments with white cabbage and also when it was made into sauerkraut. In all our grain experiments we also carried out baking tests.

Milk Processing

There are great differences when producing goats' milk butter and goats' cheese under the influence of different Moon-zodiac aspects. On Water days one needs twice as much time to produce butter as one does on Light and Warmth days, and there is a marked difference in taste. Butter, which is made on Light days, has a wonderful nutty, spicy flavour but the butter made on Water days tastes unpleasantly of goat. The same was observed when making cheese.

Index